Frontiers in Drug Safety

(*Volume 1*)

Adulteration Analysis of Some Foods and Drugs

Edited by

Alankar Shrivastava

Department of Pharmaceutical Quality Assurance,
Lloyd School of Pharmacy, Greater Noida, Knowledge Park II,
Greater Noida, Uttar Pradesh 201310, India

General:

1. Any dispute or claim arising out of or in connection with this License Agreement or the Work (including non-contractual disputes or claims) will be governed by and construed in accordance with the laws of the U.A.E. as applied in the Emirate of Dubai. Each party agrees that the courts of the Emirate of Dubai shall have exclusive jurisdiction to settle any dispute or claim arising out of or in connection with this License Agreement or the Work (including non-contractual disputes or claims).
2. Your rights under this License Agreement will automatically terminate without notice and without the need for a court order if at any point you breach any terms of this License Agreement. In no event will any delay or failure by Bentham Science Publishers in enforcing your compliance with this License Agreement constitute a waiver of any of its rights.
3. You acknowledge that you have read this License Agreement, and agree to be bound by its terms and conditions. To the extent that any other terms and conditions presented on any website of Bentham Science Publishers conflict with, or are inconsistent with, the terms and conditions set out in this License Agreement, you acknowledge that the terms and conditions set out in this License Agreement shall prevail.

Bentham Science Publishers Ltd.
Executive Suite Y - 2
PO Box 7917, Saif Zone
Sharjah, U.A.E.
Email: subscriptions@benthamscience.org

BENTHAM SCIENCE

CONTENTS

FOREWORD

This is a wonderful book on the one of the topic which is directly related to the health of the people. The topic of the book is self explanatory "*Adulteration Analysis of Some Foods and Drugs*". The information related to adulteration analysis is now become one of the fundamental right of the people. This is well known fact that adulteration of foods and drugs is common mishaps in the part of world which is either not developed or facing the challenges to be count in the developed countries. The adulteration of colorants, aphrodisiacs, herbal supplements and dairy products are not rare. The concept of the book is straight and the technical depth is kept balanced to be understood by the young scientific community as well as researchers of this field.

Major food adulteration and contamination events seem to occur with some regularity, such as the widely publicised adulteration of milk products with melamine and the recent microbial contamination of vegetables across Europe for example. With globalization and rapid distribution systems, these can have international impacts with far-reaching and sometimes lethal consequences [1].

There are number of rules and regulations related to mixing anything which should be not present in any food or drugs. The intention of such mixing is very clear; to increase the profit. Adulteration is thus one of the serious crimes for the mankind.

Adulteration is a major concern for both the food industry and consumers for many reasons. Thus the compendia containing analytical methodologies for the determination of adulterants is one of the needs in the current scenario.

The technical depth and knowledge covered by the book delighted me and I could not simply stop myself to attend to it.

I think the authors will found it one of the interesting guide of this field and in all I think you will find this book engaging in every chapter. I feel this will be a tonic stimulant for the flow of knowledge in its field.

REFERENCES

[1] Ellis DI, Brewster VL, Dunn WB, Allwood JW, Golovanov AP, Goodacre R. Fingerprinting food: current technologies for the detection of food adulteration and contamination. Chem Soc Rev 2012; 41: 5706-27.
[http://dx.doi.org/10.1039/c2cs35138b]

Shushma Talegonkar
Jamia Hamdard University
New Delhi
India
E mail: stalegaonkar@jamiahamdard.ac.in

PREFACE

"The food you eat can be either the safest and most powerful form of medicine or the slowest form of poison"

Ann Wigmore

The adulteration of food and drug is one of the emerging problem especially in developing and underdeveloped countries.

Medicinal products are highly regulated to ensure high quality and safety. Of particular importance is the identification and quantification of pharmaceutical impurities in medicinal products, active ingredients and pharmaceutical excipients, as they may influence the efficacy and safety of pharmaceutical products even in small amounts [1,2].

The adulteration of food products is of primary concern for consumers, food processors, regulatory agencies, and industries [3].

The adulteration and Substitution of the herbal drugs is the burning problem in herbal industry and it has caused a major advancement in the research on commercial natural products. The deforestation and extinction of many species and incorrect identification of many plants has resulted in adulteration and substitution of raw drugs [4].

No doubt, substitution is helpful in places where unavailability of particular crude drug and or unwanted adverse effects of desired crude drug are there and have a choice of other drug with similar pharmacological effect and less unwanted after effects. But in most cases, it is unacceptable because the conversion of authentic drug into substandard drug may cause variety of adverse effects from mild and moderate to severe life threatening reactions [5].

Herbal adulteration is one of the common malpractices in herbal raw material trade [6].

The adulteration of dietary supplements has been reported fairly frequently, yet most consumers and health care professionals are unfamiliar with the problem [7].

Thus there if the food or drug is found to be adulterated the analysis of adulterants is important. This forms the basis for which this project is conceived and compiled in to the form of book.

The first chapter of the book (Authors: Shrivastava A & Kumar R) is an introductory portion. The following chapter deals with some basics related to the topic of book like definitions and explanations like food, food fraud history of food adulteration and ends up with some brief description of *Codex Alimentarius*.

The second chapter (Authors: Agrimonti C & Marmiroli N) reviews the application of genomics against adulteration in dairy food chain, in particular application to identify milk and cheese animal origin. Amplification of single traits of animal DNA, residual in milk and cheese, through end point or real time PCR, allowed the identification, and in some cases the quantitation, of animal contribute to dairy products.

The third chapter (Authors: Fang M, Chia-Fen T, & Hwei-Fang C) is about inspection of colorant adulteration by modern LC mass spectrometry. This chapter introduces modern mass spectrometry for the detection of dyes in foods. This chapter also includes the application of DIA (data-independent acquisition) for simultaneously screening and confirmation of dyes in various foods.

The fourth chapter (Author: Shrivastava A) is about analysis of adulteration in Honey. This chapter describes briefly nutritious benefits and composition of honey. The analysis of adulteration of honey is described in different sections; Liquid Chromatography (LC), Nuclear Magnetic Resonance (NMR), Near Infrared Spectroscopy (NIS), Rheology and Sensors.

The fifth chapter (Author: Shrivastava A) is about adulteration analysis of drugs by UV spectrophotometry. UV Vis spectrophotometry is one of the basic analytical technique in the analysis of many inorganic and organic molecules. It is also used for the determination of complex mixtures to some extent. UV Vis spectrophotometry is used for both industrial and academic research. This chapter includes determination of adulteration of some drugs by using this versatile technique.

The sixth chapter (Authors: Özge T & Taner B) described the adulteration analysis of Pomegranate Juice. An adulterated pomegranate juice can be identified if its chemical composition differs from the range of a pure juice. The most frequently methods used to detect adulteration of pomegranate juice are based on the profiling and quantification of a number of compounds such as carbohydrates, phenolic compounds, amino acids, anthocyanins and pigments, and organic acids. Traditional chemical analysis techniques such as high performance liquid chromatography (HPLC), gas chromatography (GC), and attenuated total reflection (ATR)-Fourier transform infrared (FTIR) spectroscopy have also been successfully used to determine the authenticity of pomegranate juices.

The seventh chapter (Authors: Schramek N & Wollein U) is about adulterations in Dietary Supplements. This chapter highlights the most common synthetic substances found as adulterants in dietary supplements, and describes the analysis and the difficulties with the scientific assessment, using sexual enhancement products as an example.

The chapter eight (Authors: Demirezer LO & Ucakturk E) is about adulterations of food supplements and analytical methodologies for their quality control. The adulteration or substitution of the herbal products especially food supplements are the major problem in herbal industry and it has caused a major treat in the research on commercial natural products. Thin-layer chromatography, liquid chromatography, gas chromatography and capillary electrophoresis, chromatographic fingerprinting analysis are employed for the identification and quantification of bioactive compounds, impurities, contaminants, and also other compounds present in the food supplements.

The chapter nine (Authors: Sherazi STH, Mahesar SA, Kandhro AA, & Sirajuddin) is about *trans* fat analysis in some foods. *trans* fat is supposed to be unwanted fat and in the diet it should be as minimum as possible. Therefore, labeling of *trans* fat is mandatory in all foods in which fats and oils are involved. For the labeling of *trans* fat on food products adequate analytical methods are required. Therefore, the present review will cover brief outline of analytical techniques used for the analysis of *trans* fat.

The last and tenth chapter (Author: Sharma B) is about analysis of adulterants in Saffron. Saffron is of commercial importance for manufacturers because of its flavoring properties in food industry and medicinal properties in pharmaceutical industry. The biological source is the dried stigma of the flower *Crocus sativus* L. In the present chapter, adulteration of saffron and various physical, chemical and analytical techniques like TLC and spectroscopy etc. for the rapid detection and identification of pure and fake saffron are described.

I hope this book will extend learning beyond the limitations of some websites and articles published in the segment of adulteration analysis in some foods, dietary supplements and drugs. I welcome the authors to send their suggestions via email.

REFERENCES

[1] Schramek N, Wollein U, Eisenreich W. Pyrazolopyrimidines in 'all-natural' products for erectile dysfunction treatment: the unreliable quality of dietary supplements. Food Addit Contam Part A Chem Anal Control Expo Risk Assess 2015; 32(2): 127-40.
[http://dx.doi.org/10.1080/19440049.2014.992980]

[2] Roy J. Pharmaceutical impurities–a mini-review. AAPS PharmSciTech 2002; 3: 1-8.

[3] Kalivas JH, Georgiou CA, Moira M, Tsafaras I, Petrakis EA, Mousdis GA. Food adulteration analysis without laboratory prepared or determined reference food adulterant values. Food Chem 2014; 148: 289-93.
[http://dx.doi.org/10.1016/j.foodchem.2013.10.065]

[4] Prakash O. Jyoti, Kumar A, Kumar P. Niranjan Kumar MannaAdulteration and Substitution in Indian Medicinal Plants: An overview. Journal of Medicinal Plants Studies 2013; 1(4): 127-32.

[5] Ahmed S, Hasan MM. Crude Drug Adulteration: A Concise Review. World J Pharm Pharm Sci 4(10): 274-83.

[6] Sagar PK. Adulteration and substitution in endangered ASU medicinal plants of India: A Review. Int J Med Arom Plants 2014; 4(1): 56-73.

[7] Cole MR, Fetrow CW. Adulteration of Dietary Supplements. Am J Health Syst Pharm 2003; 60(15)http://www.medscape.com/viewarticle/460319

Alankar Shrivastava
Department of Pharmaceutical Quality Assurance
Lloyd School of Pharmacy Greater Noida, Knowledge Park II
Greater Noida, Uttar Pradesh 201310
India
E mail: alankarshrivastava@gmail.com

List of Contributors

Alankar Shrivastava	Department of Pharmaceutical Quality Assurance, Lloyd School of Pharmacy, Knowledge Park II, Greater Noida, Uttar Pradesh 201310, India
Ritesh Kumar	Department of Pharmaceutics, Institute of Biomedical Education and Research, Mangalayatan University, Aligarh 202146, India
Caterina Agrimonti	Department of Chemistry, Life Sciences and Environmental Sustainability, University of Parma , V.le Parco Area delle Scienze 11/A 43124 Parma, Italy
Nelson Marmiroli	Department of Chemistry, Life Sciences and Environmental Sustainability, University of Parma , V.le Parco Area delle Scienze 11/A 43124 Parma, Italy
Mingchih Fang	Division of Research and Analysis, Taiwan Food and Drug Administration, Taipei City, Taiwan
Chia-Fen Tsai	Division of Research and Analysis, Taiwan Food and Drug Administration, Taipei City, Taiwan
Hwei-Fang Cheng	Division of Research and Analysis, Taiwan Food and Drug Administration, Taipei City, Taiwan
Alankar Shrivastava	Department of Pharmaceutical Quality Assurance, Lloyd School of Pharmacy, Greater Noida, Knowledge Park II, Greater Noida, Uttar Pradesh 201310, India
Özge Taştan	Ege University, Faculty of Engineering, Department of Food Engineering, 35100 Bornova, Izmir, Turkey
Taner Baysal	Ege University, Faculty of Engineering, Department of Food Engineering, 35100 Bornova, Izmir, Turkey
Nicholas Schramek	Bavarian Health and Food Safety Authority, Oberschleißheim, Germany
Uwe Wollein	Bavarian Health and Food Safety Authority, Oberschleißheim, Germany
L.O. Demirezer	Hacettepe University, Faculty of Pharmacy, Dept. of Pharmacognosy, 06100 Ankara, Turkey
E. Ucakturk	Hacettepe University, Faculty of Pharmacy, Dept. of Analytical Chemistry, 06100 Ankara, Turkey
Syed Tufail Hussain Sherazi	National Centre of Excellence in Analytical Chemistry, University of Sindh, Jamshoro-76080, Pakistan
Sarfaraz Ahmed Mahesar	National Centre of Excellence in Analytical Chemistry, University of Sindh, Jamshoro-76080, Pakistan
Aftab Ahmed Kandhro	Dr. M. A. Kazi Institute of Chemistry, University of Sindh, Jamshoro-76080, Pakistan
Sirajuddin	National Centre of Excellence in Analytical Chemistry, University of Sindh, Jamshoro-76080, Pakistan
Brijesh Sharma	Institute of Biomedical and Education Research, Mangalayatan University, Beswan, Aligarh, 202146, India

CHAPTER 1

Introduction (Food, Food Fraud, History of Adulteration)

Alankar Shrivastava[1,*] and **Ritesh Kumar**[2]

[1] *Department of Pharmaceutical Quality Assurance, Lloyd School of Pharmacy, Knowledge Park II, Greater Noida, Uttar Pradesh 201310, India*

[2] *Department of Pharmaceutics, Institute of Biomedical Education and Research, Mangalayatan University, Aligarh 202146, India*

Abstract: Food adulteration is not new or unknown thing for the people and is now one of the global threats for the mankind. The following chapter deals with some basics related to the topic of book like definitions and explanations like food, food fraud history of food adulteration and ends up with some brief description of Codex Alimentarius. This chapter will increase the common understanding of the readers to proceed in the next important chapters.

Keywords: Adulteration, Aesthetic adulteration, Codex Alimentarius, Economic adulteration, Food, Food fraud, Food adulteration, Food safety, Hazardous adulteration, History of adulteration, Safe food.

INTRODUCTION

Someone has rightly said that *"We are what we eat."* The food that we eat in day today life must be utilized, transformed into different forms, and/or removed by our bodies through excretion process. Food is one of the essential component for life to sustain in this planet, and its access is often the limiting factor in the size of a given population [1].

An important function of the food is to maintain the supply of all macronutrient (*e.g.* fat, protein, carbohydrate) and micronutrient (*e.g.* vitamins and minerals) to the body. As this is one of the essential component, the organoleptic properties and safety aspects are also important. Thus food safety is of utmost importance. The food we consume should be:

* **Corresponding author Alankar Shrivastava:** Department of Pharmaceutical Quality Assurance, Lloyd School of Pharmacy, Knowledge Park II, Greater Noida, Uttar Pradesh 201310, India; Tel: +91-7351002560; E-mail: alankarshrivastava@gmail.com

1. free from any kind of toxins.
2. free from any kind of contaminants.
3. free from any kind of pathogenic microorganisms.
4. free from any food-poisoning microorganism.
5. there should be balance of nutrients as per our needs [2].

Food

Food is essential for life and is required not only for sustaining growth but also for combating diseases. All living things are nourished by macro and micro nutrients of food and some foods are known to have health benefits. Foods with health benefits or foods that modulate biological functions of the body and aid prevention and/or relief from pathological states are commonly known as nutraceuticals or functional foods or dietary supplements [3].

There is huge varieties of items and product which can be termed as 'food'. Food is defined as something that provides pleasure and nutrition when eaten. There are many forms and varieties of food. These may be classified under three types, (1) Fresh and raw products, *e.g.*, meat, fish, fruits and vegetables (fresh) (2) Dairy Products, *e.g.*, yogurt, cheese, butter and milk *etc*. (3) Cooked or baked foods, *e.g.*, biscuits, prepared meals and bread. Nowadays the demand of ready-prepared food and their components is also increased in addition with the supply of basic food materials [4].

Food Fraud

S. Walter (2013) [5] published the statement of John Spink, associate director of the Anti-Counterfeiting and Product Protection Program at Michigan State University.

"There's probably no more fraud per capita or per person today than there was in ancient Roman times. There's just more people now. And it's multiplied because of globalization and manufacturing."

Economic gain is the reason for the crime known as 'Food fraud'. This is an act of defrauding consumers or food manufacturers, retailers or importers. Most of the time consumers failed to notice the quality of items they have purchased and thus left undetected. Sometimes adulterated food does not produce any harmful effects and forms another reason for undetected cases. This act defamed the food industry throughout history [6].

The fraud related to food is considered as prime concern related to the economy. Any change in the purity and/or identity of food material/product and its ingredients by altering, replacement or dilution by chemical or physical means results in the adulteration. The types of "food frauds" were categorized into three categories:

1. Replacement,
2. Addition or
3. Removal

The database related to food fraud and adulteration was firstly developed by USP. The parties interested in some specific products can assess the threats by the information provided in this database [7]. According to this database published in 2012, ninety-five percent of the cases were related to replacement of original material partially or wholly by cheaper substitute less expensive. For example, hazelnut oil is cheaper than olive oil and thus former is used to substitute latter to increase profits. Substitution of Chinese star anise with Japanese star anise (harmful), replacement of high-quality spices with low quality spices are some more examples of such kinds of adulterations. There are reports describing mixing lead chromate/tetraoxide in low quality spices to emanate color of high quality spices is one of the methods of food fraud [8].

The fraud may cost between ten billons and fifteen billion per year, according to the Grocery Manufacturers Association. Approximately this affects the sale of 10% of commercial food products. Both financial and public relations of any company or industry related to food are affected by such kind of food frauds [9].

The expectation of safe and suitable food is a people's right. The injury caused by unsafe food is of course unpleasant and in the worst case, may be fatal. This may also have impact on economy by damaging tourism and trade which may be precipitated into unemployment and litigation [10].

In the past five decades the capacity of the world for providing food to the people is increased because of utilization of technology. Now the productivity and diversity of food increased with less dependence on the environmental conditions [11]. With the boom in the economy of many countries the exchange of food whether may be by barter or sell is increased, and there have been concern about the quality and safety of the exchanged food [12].

Food Safety

There are many low and middle income countries in the world where regulatory, surveillance, and control systems are unable to address the range of potential hazards [13]. Over the last two decades, food safety rules have increased considerably, both in numbers and in scope. Some of these rules are formed by governments at the international, national and regional level. Others are drawn up by private enterprises or associations of private enterprises [14].

As per WHO's estimates, each year there are more than 1000 million cases of acute diarrhea in children below five years in developing countries (WHO, 2008). The Chinese report of death of at least six babies and about 0.3 million infants suffered from urinary problem and kidney stones after drinking milk or infant formula because of contamination of melamine in December 2008. There are incidents in India too such as children have fallen sick after eating contaminated midday meals, adulteration of Kuttu flour in festival season. Food borne illness can damage trade and tourism, and will lead to loss of earning, unemployment and litigation [15].

Anklam and Battaglia (2001) in their publication described about few adulterations of last two decades of nineteenth century. Wine was adulterated with diethyglycol, sodium azide and saccharin for sooth taste, prevention of microbial spoilage and to make it sweet respectively. Honey was adulterated with high fructose syrup and bromo-acetic acid added to the beer for microbial growth prevention. Maple syrup and fruit juices were diluted with water and starch hydrolysates. Some other examples are springbok meat labelled as venison, as well as salmon-trout labelled as salmon, and whitefish labelled as perch [16].

One of the recent cases is identification of lead in and monosodium glutamate in the popular instant food, Maggi noodles in India. This only reflects the overall lack of standards in food products in the country. There is a heightened concern in the case of Maggi because it is the favorite food of young children [17].

The world's population has been increasing rapidly since the beginning of the 20th century, from about 1.6 billion in 1900, 2.5 billion in 1950, 6.1 billion in 2000, to 7.0 billion in 2011, according to estimates by the United Nations Population Fund [18]. The world population could grow to about 16.5 billion by the end of the century [19].

Thus there is no doubt that there will be huge increase in the per capita kcal consumption in the future. The expenditure on food is major part of income in developing countries thus they want to buy maximum in minimum price possible. This attitude of consumers motivates traders or manufacturers for adulteration [20].

Gupta and Panchal (2009) [21] described in their publication about types of adulteration and examples. This is described under (Table 1). This paper also described reasons of the adulteration; (a) Availability of many products (b) Psychology of the consumer (c) Too much bargain (d) Wrong buying practice of consumer (e) Availability of adulterants. In addition to this, practice of adulteration also increased if demand exceeds the supply of any product. Various reasons of adulteration are presented under (Fig. **1**).

Table 1. Different types of adulteration and examples [21].

Types of adulteration		
Intentional adulterants	*Metallic contamination*	*Incidental adulterants*
Examples: sand, marble chips, stone, mud, chalk powder, water, mineral oil and coal tar dyes	Examples: arsenic from pesticides, lead from water, and mercury from effluents of chemical industries, tin from cans *etc*.	Examples: Pesticide residues, tin from can droppings of rodents, larvae in foods. Metallic contamination with arsenic lead, mercury

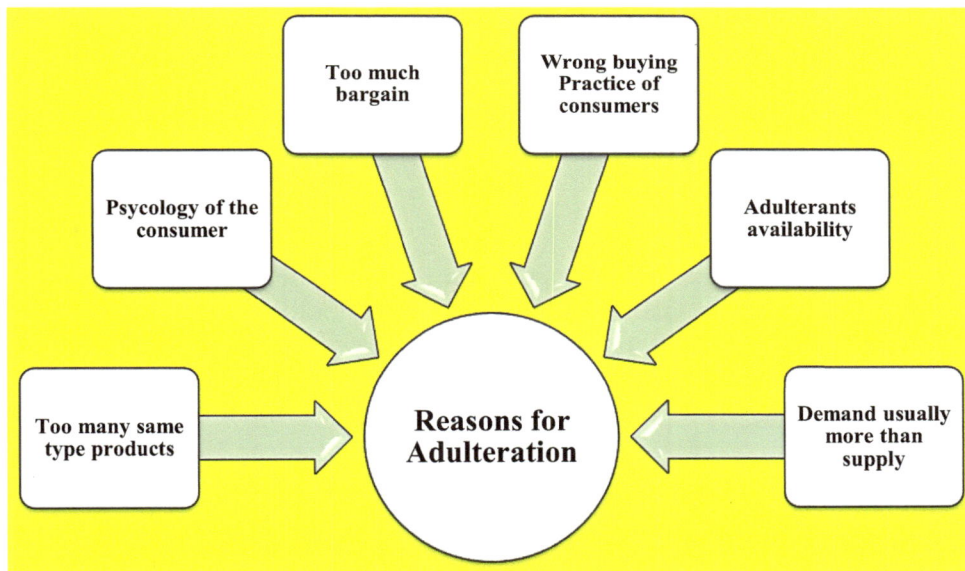

Fig. (1). Various reasons of adulteration.

Food Adulteration

Food adulteration and fraud and attempts at their control have a long history [22].

Food adulteration provisions of the Food, Drug, and Cosmetic Act govern three types of adulteration: economic, aesthetic, and hazardous.

Economic adulteration is the oldest form of consumer fraud. Watered milk, for example, is not a health problem, but it is an economic problem and is prohibited by law. Aesthetic adulteration involves more analysis. Under the FD&C Act, food that contains any filthy or decomposed matter, or that is 'otherwise unfit for food', is adulterated and therefore illegal. This is true, even if the food is absolutely sterile and presents no health problem. These aesthetic provisions in the law protect the consumer's sensitivity and health.

Any food held under unsanitary conditions that may present a danger to health is adulterated and illegal. The FDA has taken most of its enforcement action under the 'unsanitary conditions' provision of the food law, the backbone of food and drug protection in the United States [23].

The popular idea of adulteration is that it always consists of the addition of cheaper ingredients, often harmful, to foods, beverages, drugs, confectionery, and other commodities. Very likely this was the original form of adulteration and it is often practiced, but it comes very far from comprising the whole meaning of the word [24]. Food authenticity has become a focal point attracting the attention of producers, consumers, and policy makers. A range of analytical methods to detect fraud has to be developed, modified, and reappraised on a continuous basis to be a step ahead of those pursuing these illegal activities [25].

History of Adulteration

In ancient Rome and Greece, wine was often mixed with honey, herbs, spices and even saltwater, chalk or lead—which served as both a sweetener and a preservative. After this, adulteration with food grows gradually for economic gain. Spices were very valuable during middle ages and thus targeted for adulteration by mixing with ground nutshells, pits, seeds, juniper berries, stones or dust [26].

Mesopotamian, Hebrew, Egyptian, Greek and Roman authorities had provisions on good practices, inspectors for the integrity of precious commodities or goods (Grains, wines *etc.*), and penalties and punishments for those offended the public interest and often jeopardized the health of citizens [27].

Many descriptions are available showing the attempts to prevent food frauds by various rules and regulations in our ancient history. For example, India has regulations for prohibiting adulteration of grains and edible fats in more than 2000 years back. Mosaic and Egyptian laws, about 2500 years ago had provisions to prevent the contamination of meat [28].

In the middle ages food quality was mostly related to weight and quality of food stuffs. Adulterations occurred in the food stuffs which are highly valuable or the part of the food of major population or which are traded in large quantities. Since during this time period spices were the one of the valuable component that has be traded across the borders of continents even, are prone for adulteration. Governments of that time also were sensitive to prevent their population for any kind of such food fraud. For example, is passing of a law; *Assisa Panis et Cervisae* (Assize of Bread and Ale) in England in year 1266 for the regulation of weight and price of bread and ale in relation with corn. Some documents also show the appointment of food inspectors to stop adulteration of food stuffs that had a potential of endangering public health [29].

Penalties in olden times were quite harsh. In Salerno, Italy in the 10th century, the penalty for adulteration of a drug included "whoever shall have or sell any noxious drug or poison not useful or necessary to his art, let him be hanged" [30].

In Nuremberg in the fifteenth century, an adulterator of saffron was burnt over his own produce; others were buried alive or their eyes were gouged out. Other forms of punishment comprised expulsion, whipping, cutting off ears, and drowning. In some cases, offenders were forced to consume their adulterated food until they died. Later, French King Louis XIV, the "Sun King", imposed capital punishment for the adulteration of wine by the addition of pokeweed (*Phytolacca americana* L.) [31].

Between the 13th and 16th centuries, bread, wine, beer, spices, and valuable natural coloring materials were often adulterated. In England in 1319, a meat market overseer succeeded in putting a butcher in the pillory for selling unsound beef [32].

Frederick Accum was the first to raise the alarm about food adulteration. Accum was a German chemist who had come to London in 1793 and who quickly established himself as a chemical analyst, consultant and teacher of chemistry [33].

Accum in 1820, published a book title "A Treatise on Adulterations of Food and Culinary Poisons". Each section details the methods of adulteration and how they may be detected. The largest section is devoted to the adulteration of beer. The first edition of a thousand copies was sold out within a month of publication and the second edition appeared later in the same year [34].

Harvey W. Wiley, known as the "Father of the Pure Food and Drugs Act" of United States [35], was chief chemist of the Agriculture Department's Bureau of Chemistry. He investigated the safety of some preservatives used in food. Wiley recruited volunteers, which the press soon dubbed the Poison Squad. These volunteers ate foods containing measured amounts of borax, salicylic acid, formaldehyde, and other chemical preservatives [36].

The initial five preservatives studied were borax, salicylic acid, sulfuric acid, sodium benzoate, and formaldehyde. Dosages ranged from one-half gram daily to four grams by the end of the five-year study. Wiley stopped the experiments only when the chemicals made several of the diners so sick that they couldn't function--nausea, vomiting, stomachaches, and the inability to perform work of any kind. This is because of his study Congress, in 1906, passed both the Meat Inspection Act and the original Food and Drugs Act, prohibiting the manufacture and interstate shipment of adulterated and misbranded foods and drugs [37].

During the Industrial Revolution at the turn of the 19th Century, people moved towards towns and cities for work and became reliant on food retailers as a result. Industrial managers became increasingly infuriated by high levels of absenteeism, a major cause of which was the consumption of adulterated foods [38].

By the second half of the nineteenth century, food adulteration had become so widespread that governments were forced to act to preserve public order. The food adulteration in industrial England recounting the contamination of staple foods: vitriol in beer, cyanide in wine, brisk dust in cocoa, and alum in white bread, that led to widespread public concern as well as implementation of the first comprehensive food safety regulations and foundation of the discipline of the food science [39].

Some more examples of adulterations in nineteenth century are vinegar with sulphuric acid, pickles colored with copper, sugary confections dyed red with lead and pepper mixed with floor sweepings, coffee was diluted with chicory, acorns, or a type of beer called mangelwurzel [40].

The need for improved health and food control and the rapidly expanding international food trade stimulated cooperation on an international level. After World War II, activity in international standardization started intensively in the framework of ISO. A Joint FAO/WHO Food Standards Program was established in 1962, and a joint subsidiary body was created: the Codex Alimentarius Commission (CAC). The trend in the field of food regulation is characterized by growing efforts for harmonization at an international level [41].

The first general law against food adulteration in the United States was enacted in Massachusetts in 1784; gradually, other states passed a variety of food and drug statutes. As the country expanded, however, it became clear that a national law was needed. From 1879 to 1906 more than 100 food and drug bills were introduced in the U.S Congress. Later on Congress passed the first national legislation designed to control impure and unsafe food and drugs: The pure and food drug (Wiley) act of 1906 [42].

Codex Alimentarius

With the purpose of protecting the health of consumers and ensuring fair practices in the food trade Food and Food and Agriculture Organization of the United Nations (FAO) and the World Health Organization (WHO) jointly established in the year 1962 an intergovernmental body of over 180 members know as ***Codex Alimentarius Commission*** (CAC). The ***Codex Alimentarius*** (Latin, meaning Food Law or Code) is the result of the Commission's work: a collection of internationally adopted food standards, guidelines, codes of practice and other recommendations [43].

Some other aims were (1) to promote coordination of all food standards work undertaken by international governmental and non-governmental organizations; (2) to determine priorities and initiate and guide the preparation of draft standards through and with the aid of appropriate organizations; and (3) to finalize standards [44].

The Codex Alimentarius includes provisions in respect of food hygiene, food additives, residues of pesticides and veterinary drugs, contaminants, labelling and presentation, methods of analysis and sampling, and import and export inspection and certification [45].

CONCLUSION

In this way some common fundamentals of the topics were explained here. This chapter forms the basis for the readers to better understand the other chapters. The motto of writing this chapter is of course to initiate the topic for which this literature is produced. In the next chapter authors will discuss in detail about various adulteration analysis methods of some foods and drugs.

CONSENT FOR PUBLICATION

Not applicable.

CONFLICT OF INTEREST

The author (editor) declares no conflict of interest, financial or otherwise.

ACKNOWLEDGNEMT

Declared none.

REFERENCES

[1]   Lehotay SJ. Hajs˘lova´ J. Application of gas chromatography in food analysis. Trends Analyt Chem

2002; 21(9+10): 686-97.
[http://dx.doi.org/10.1016/S0165-9936(02)00805-1]

[2]   Patrick F. Fox Introduction to Analysis in the Dairy Industry Handbook of Dairy Foods Analysis. CRC Press 2010; p. 1.

[3]   Krishnaraju AV, Bhupathiraju K, Sengupta K, Golakoti T. Regulations on Nutraceuticals, Functional Foods and Dietary Supplements in India. Nutraceutical and Functional Food Regulations in the United States and Around the World. DOI: 2014, Elsevier Inc.

[4]   MacDougal DB. Colour in food: Improving quality. Woodhead Publishing Limited, England, pp. 2.

[5]   Walter S. Farm Fakes: A History of Fraudulent Food 2013.http://modernfarmer.com/ 2013/05/farm-fakes-a-history-of-fraudulent-food/

[6]   Johnson R. Food fraud and "economically motivated adulteration" of food and food ingredients. In food fraud and adulterated ingredients: background, issues, and federal action. Braden DT, 2014 by Nova Science Publishers, Inc. pp. 1-2.

[7]   US Pharmacopeia. New research reveals food ingredients most prone to fraudulent economically motivated adulteration http://www.eurekalert.org/pub_releases/2012-04/up-nrr040512.php

[8]   Moore JC, Spink J, Lipp M. Development and application of a database of food ingredient fraud and economically motivated adulteration from 1980 to 2010. J Food Sci 2012; 77(4): R118-26.
[http://dx.doi.org/10.1111/j.1750-3841.2012.02657.x] [PMID: 22486545]

[9]   Johnson R. Johnson R. Food Fraud and "Economically Motivated Adulteration" of Food and Food Ingredients. January 10, 2014. Available online: https://www.fas.org/sgp/crs/misc/R43358.pdf

[10]   Indian Standard. 2012.https://law.resource.org/pub/in/bis/S06/is.16020.2012.pdf

[11]   Stearns DJD. Food safety and the law: Understanding the real life liability risk.Food Safety: Researching the Hazard in Hazardous Foods. Torronto, New Jersey: Apple Academic Press 2014; p. 124.

[12]   Unnevehr L, Hoffmann V. Food safety management and regulation: International experiences and lessons for China. J Integr Agric 2015; 14(11): 2218-30.
[http://dx.doi.org/10.1016/S2095-3119(15)61112-1]

[13]   Safety F. Global Panel on agriculture and food system for nutrition 2016.http://glopan.org/food-safety

[14]   Gibbon P. Stefeno Ponte, Evelyne Lazaro Global Agro-Food Trade and Standards: Challenges for Africa. Palgrave Macmillan 2010; p. 224.
[http://dx.doi.org/10.1057/9780230281356]

[15]   Shukla S, Shankar R. S.P. Singh. Food safety regulatory model in India. Food Control 2014; 37: 401-13.
[http://dx.doi.org/10.1016/j.foodcont.2013.08.015]

[16]   Anklam E, Battaglia R. Food analysis and consumer protection. Trends Food Sci Technol 2001; 12: 197-202.
[http://dx.doi.org/10.1016/S0924-2244(01)00071-1]

[17]   Widespread food adulteration, Free Press Journal. Jun 06, 2015. Available online: http://www.freepressjournal.in/widespread-food-adulteration/597989

[18]   Chen J, Shi H, Sivakumar B, Peart MR. Population, water, food, energy and dams. Renew Sustain Energy Rev 2016; 56: 18-28.
[http://dx.doi.org/10.1016/j.rser.2015.11.043]

[19]   United Nation Population Fund. World population trends http://www.unfpa.org/ world-population-trends#sthash.qaHY2opf.dpuf

[20]   A report on Global Food and Nutrition Scenarios. Millennium Institute, Washington, DC, March 15th, 2013. Available online: http://www.un.org/en/development/desa/policy/wess/wess_bg_papers/

bp_wess2013_millennium_inst.pdf

[21]   Gupta N, Panchal P. Extent of Awareness and Food Adulteration Detection in Selected Food Items Purchased by Home Makers. Pak J Nutr 2009; 8: 660-7.
[http://dx.doi.org/10.3923/pjn.2009.660.667]

[22]   Deelstra H, Thorburn Burns D, Walker MJ. The adulteration of food, lessons from the past, with reference to butter margarine and fraud. Eur Food Res Technol 2014; 239: 724-44.
[http://dx.doi.org/10.1007/s00217-014-2274-0]

[23]   Featherstone S. Food regulations, standards, and labeling A Complete Course in Canning and Related Processes. Elsevier Ltd. 2015.

[24]   Bradley TJ, Boston M. What is adulteration? J Am Pharm Assoc 1913; 2(1): 53-6.
[http://dx.doi.org/10.1002/jps.3080020113]

[25]   Ulberth F, Buchgraber M. Authenticity of fats and oils. Eur J Lipid Sci Technol 2000; 102: 687-94.
[http://dx.doi.org/10.1002/1438-9312(200011)102:11<687::AID-EJLT687>3.0.CO;2-F]

[26]   Schumm L, Fraud F. A Brief History of the Adulteration of Food 2014.http://www.history.com/news/hungry-history/food-fraud-a-brief-history-of-the-adulteration-of-food

[27]   Tsimidou MZ, Ordoudi SA, Nenadis N, Mourtzinos I. Food Fraud, Encyclopedia of Food and Health. Elsevier Limited 2016; p. 35.
[http://dx.doi.org/10.1016/B978-0-12-384947-2.00010-6]

[28]   Ihegwuagu Nnemeka Edith, Emeje Martins Ochubiojo. Food Quality Control: History, Present and Future. Scientific, Health and Social Aspects of the Food Industry. Available from: http://cdn.intechweb.org/pdfs/27397.pdf

[29]   Melitta Weiss Adamson. Food in Medieval Times. London, Connecticut: Greenwood Press Westport 2004; p. 64.

[30]   Drug Safety News: Counterfeit Medicines and Adulteration. 2011 Published on British Columbia Drug and Poison Information Centre (BC DPIC). Available on: http://www.dpic.org/article/professional/drug-safety-news-counterfeit-medicines-and-adulteration

[31]   Schieber A. Introduction to Food Authentication. Modern techniques for food authentication. Editor: Da-Wen Sun. First edition 2008, Academic Press is an imprint of Elsevier. pp. 2.

[32]   Tsimidou M, Boskou D. Adulteration of foods, History and Occurrence Encyclopedia of Food Sciences and Nutrition. 2nd ed. 2003; pp. 42-7.
[http://dx.doi.org/10.1016/B0-12-227055-X/00012-2]

[33]   The fight against food adulteration http://www.rsc.org/Education/EiC/issues/2005Mar/Thefightagainstfoodadulteration.asp

[34]   Frederick Carl Accum. http://www.rsc.org/images/cw01_accum_tcm18-196603.pdf

[35]   Harvey W. Wiley. FDA Consumer magazine, The Centennial Edition / January-February 2006. Available   from: http://www.fda.gov/aboutfda/whatwedo/history/centennialoffda/harveyw.wiley/default.htm

[36]   Meadows M. A Century of Ensuring Safe Foods and Cosmetics 2006.http://www.fda.gov/AboutFDA/WhatWeDo/History/FOrgsHistory/CFSAN/ucm083863.htm

[37]   Lewis Carol. The "Poison Squad" and the Advent of Food and Drug Regulation. U.S. Food and Drug Administration Consumer Magazine November-December, 2002, pp. 1.

[38]   Griffiths S. History of UK food law http://fstjournal.org/features/history-uk-food-law

[39]   Campbell H. Spurlock's vomit and visible food utopias Food Utopias: Reimagining Citizenship, Ethics and Community. 198.

[40]   Michael T. Roberts Food Law in the United States. Cambridge university press 2016; p. 40.

[41]   Lásztity R, Petró-Turza M, Földesi T. History of food quality standards Encyclopedia of Life Support Systems (EOLSS), Developed under the Auspices of the UNESCO. Oxford, UK: Eolss Publishers 2004; pp. 1-2.

[42]   Satya D. Dubey, George Y.H. Chi, Roswitha E. Kelly. The FDA and the IND/NDA statistical review process. Statistics in the Pharmaceutical Industry, Editors: C. Ralph Buncher, Jia-Yeong Tsay, 3rd Edition. CRC Press, pp. 56.

[43]   Cosbey A. A Forced Evolution? http://www.iisd.org/pdf/forced_evolution_codex.pdf

[44]   Codex Alimentarius Commission. United States Department of Agriculture http://www.fsis.usda.gov/ wps/portal/fsis/topics/international-affairs/us-codex-alimentarius/Codex+Alimentarius+Commission

[45]   Codex Alimentarius Commission. ftp://ftp.fao.org/codex/Publications/ProcManuals/Manual_24e.pdf 2015.

CHAPTER 2

Genomics for Detecting Adulteration in Dairy Food Chain

Caterina Agrimonti and **Nelson Marmiroli**[*]

Department of Chemistry, Life Sciences and Environmental Sustainability, University of Parma, V.le Parco Area delle Scienze 11/A, 43124 Parma, Italy

Abstract: Genomic platforms, improved in the last decade at high throughput level, are becoming an elective tool in food analysis, in addition to the traditional microbiological, chemical and physical methods.

Because milk and its products are among the most frequent causes of food-allergies, and cow milk is the major culprit, determination of animal origin of milk employed in a dairy manufacturing, is important to establish the safety of a supply chain.

Since 1992, the European Union has introduced labels of origin (Protected Designation of Origin, Protected Geographical Indication, Traditional Speciality Guaranteed) to eliminate unfair competition and misleading of consumers by non-genuine products. Identification of animal species in milk food chain is therefore important to protect products subjected to labeling, as buffalo Mozzarella, sheep and goat cheese or Parmesan cheese.

Amplification of single traits of animal DNA, residual in milk and cheese, through end point or real time PCR, allowed identification, and in some cases quantitation, of animal source of dairy products.

This chapter reviews application of genomics against adulteration in dairy food chain, in particular to identify animal origin of milk and cheese.

Keywords: Adulteration, Buffalo milk, Cheese, Cow milk, Dairy, DNA analysis, End point PCR, Food chain, Goat milk, Limit of detection, Limit of quantitation, Linear dynamic range, Milk, Multiplex PCR, Quantitative PCR, Real time PCR, Sheep milk.

[*] **Corresponding author Nelson Marmiroli:** Department of Chemistry, Life Sciences and Environmental Sustainability, University of Parma, V.le Parco Area delle Scienze 11/A, 43124 Parma, Italy; Tel: +39 0521-905606; E-mail: nelson.marmiroli@unipr.it

Alankar Shrivastava (Ed.)

INTRODUCTION

Labelling [1] of food content is instrumental to protect both consumers and producers from fraud, but mislabelling still remains an issue. Consumers worldwide are being more aware of the quality of their food, also because fraud and adulteration are being continuously discovered. Health worldwide crises, as in the cases of "mad cow", avian flu, dioxin in eggs, in chicken and pig meat, have prompted the need for more controls. These accidents did not constitute only a threat to health but had also serious economic consequences on the food industry. In particular, the economic damage may be exceedingly high for prized goods, with high organoleptic and healthy properties. Products strictly related to a specific geographical area and/or to a preparation method, preserved by strict and unchangeable rules, are in the front-wall of the war to fraud and adulteration. For their properties and authenticity, they are particularly appealing to the consumers.

Several studies showed that milk and derived products available in retail in Italy, Spain, Portugal, the Czech Republic, Poland, Croatia, Egypt, Taiwan, China, India and Pakistan had some level of nonconformity with the labelling declaring their animal source [2].

Milk and its products are among the more frequent causes of food allergies because milk proteins, though at low concentrations, can be strong allergens [3]. Cow milk is certainly responsible for allergies and intolerance in human [4], while milk of other species, *e.g.* goat, does not cause allergic reactions with similar frequencies, but goat milk is more expensive and its mixing with the cheaper cow milk is a typical fraud, which can cause health risk to consumers and economic damage to honest producers.

Since 1992, the European Union has introduced, with the Regulation 2081/1992 [5], the labels of origin: Protected Designation of Origin (PDO), Protected Geographical Indication (PGI) and Traditional Specialties Guaranteed (TSG) to eliminate the unfair competition within producers and misleading of consumers by non-conform products. This Regulation, enforced by EU Regulation 1151/2012 [6], was gradually expanded internationally *via* bilateral agreements between EU and non EU countries, ensures that products genuinely originating in a specific region or produced according to traditions can be identified by the label.

Issued to answer an increasing demand of more transparency in the dairy food chain, the European Regulations require that producers declare the type of milk used in manufacturing [7]. Compliance with Regulations is mandatory because people allergic to cow milk should consume dairy products made only from sheep, goat, or soy milks. Absence of a proper labelling, indicating even traces of cow milk, can increase health risk for these people.

Some PDO cheeses, made with sheep or goat milk can contain a percentage of cow milk which does not exceed a fixed limit: *e.g.* in "Murazzano" cheese, made with sheep milk, cow milk should not exceed 40%, while in "Casciottad'Urbino", cow milk should be between 20 and 30%. Validation of authenticity of commercialized cheeses is a challenge not only for qualitative detection of each kind of milk, but also for its relative quantitation.

Analytical methods to prove the geographical origin and the animal source of milk are based on electrophoresis [8, 9], immunology [10], chromatography [11] and mass spectrometry [12]. These methods are very specific but frequently lack sensitivity and not always are suitable to test milk subjected to thermal treatment or other processes as rennet and acid coagulation, dehydration, fermentation, ripening and smoking.

Milk from healthy mammary glands contains a large number of somatic cells (epithelial cells and leukocytes) that are concentrated during transformation and constitute a source of DNA. In comparison with tests based on proteins, DNA analysis has an increased sensitivity and can be applied also to highly processed dairy products, treated with high temperature, pressure or stored for long periods, because DNA is more resilient to these treatments than other milk components [13]. Development of end point and real time PCR, allows amplification of very low amounts of degraded DNA. For these reasons in the last years the identification of animal source of dairy products is based on a genomic approach: its advantages, but also its limits, are reviewed in this book chapter.

DNA in Dairy Products

Extraction of DNA from food matrices in quality and quantity sufficient for PCR is a critical step: in fact, dairy products contain fats, proteins, calcium and other components that reduce efficiency of DNA extraction and amplification [14].

Methods employed in the genomic analysis of dairy products, based on commercial kits or on laboratory experience, are reported in Table **1**.

Four commercial methods, Wizard Resin® ("Wizard"), QIAampDNA Stool® Mini Kit ("Stool"), Charge Switch Forensic® DNA Purification Kit ("Forensic") and Nucleo Spin®Food ("Food"), and three methods based on sodium dodecyl sulfate (SDS), cetyltrimethylammonium bromide (CTAB) and Tween were reviewed for their efficiency on different matrices: fresh whole and skimmed milks, yogurt, butter, cooking cream and Emmental cheese [15]. "Wizard" gave the highest yield of genomic DNA from fresh whole and semi-skimmed milk, but it was less efficient on the other products (Fig. **1A**).

Table 1. Methods for DNA extraction described in papers cited in this work.

Dairy product	Method of extraction	Reference
Milk, cheese, butter, cream, yoghurt	WizardResin® (Promega, Madison, WI, USA) QIAampDNA Stool® Mini Kit (QIAGEN GmbH, Hilden, Germany Charge Switch Forensic® DNA Purification Kit (Invitrogen, Carlsbad, CA, USA) NucleoSpin®Food (Macherey–Nagel, Düren, Germany) CTAB based method SDS based method Tween based method	[15]
Fresh milk, condensed milk, Cheese	Extraction with guanidiniumisothiocyanate followed by purification with Silica Paramagnetic Particles	[17]
Cheese	Phenol/chloroform extraction followed by purification with Wizard® Genomic DNA Purification Kit (Promega)	[22]
Cheeses	DNeasy Blood and Tissue Kit (QIAGEN)	[23]
Milk	Phenol/chloroform method	[24]
Milk	Wizard DNA cleanup kit (Promega)	[28]
Cheese	Wizard DNA cleanup kit	[29]
Cheese	Isolation kit Invisorb Spin Food I (Invitek, Co., Berlin, Germany)	[27]
Cheese	Extraction with EDTA/SDS	[32]
Cheese	DNeasyTM Tissue Kit (QIAGEN)	[19]
Governing liquid of Mozzarella	Phenol/chloroform method	[26]
Cheese	DNeasy Blood and Tissue Kit (QIAGEN)	[33]
Milk, cheese, yogurt, cream, butter	Phenol/chloroform method	[2]
Cheese	DNeasyTM Tissue Kit (QIAGEN)	[34]
Cheese	Extraction with guanidium hydrochloride	[35, 36]
Milk	DNeasy Blood and Tissue Kit (QIAGEN)	[39]
Cheese	Phenol/chloroform extraction followed by purification with Nucleospin Food kit purification (Macherey-Nagel)	[44]
Milk	Wizard DNA cleanup kit (Promega)	[41]
Milk	Wizard DNA cleanup kit (Promega)	[42]
Milk Cheese	Invisorb Cell Mini Kit (Invitek, Berlin, Germany) Invisorb Tissue Mini Kit (Invitek)	[43]
Cheese	Wizard Plus Miniprep DNA purification system (Promega)	[16]
Milk, cheese, yogurt, cream, butter	CTAB based method	[45]

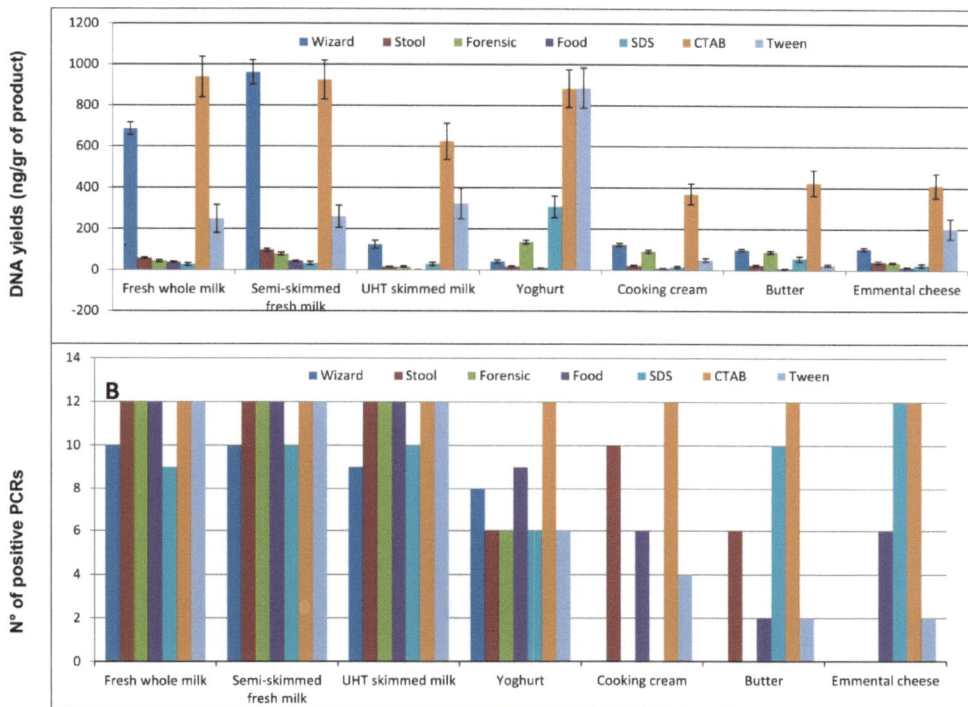

Fig. (1). Performances of different DNA extraction methods on some dairy products. **A)** mean yields expressed as ng of DNA/ gr, calculated from six independent extractions; **B)** Number of positive end point and real time PCRs, conducted on DNA extracted from dairy products with the different methods. Results derived from six end point and six real time PCR for each kind of DNA. Adapted from data of Pirondini *et al.* [15].

Satisfactory results were obtained with CTAB based method: high amounts of DNA, less degraded than those extracted with commercial kits, were obtained from all products, in particular from non-processed matrices (fresh whole and semi skimmed milk) (Fig. **1A**). Moreover, DNA extracted from all the matrices with this method was successfully amplified in all PCR tests, unlike the DNA extracted with the other methods (Fig. **1B**).

These data demonstrated that PCR amplification is not only correlated with DNA yield, but also with the absence of components, such fats or calcium ions that can inhibit DNA synthesis, or with the DNA integrity.

To optimise DNA extraction protocol, the PCR-relevant "DNA-index" was introduced [16]: it is calculated on the sum of positive PCRs conducted on DNA extracted from cheeses at different time of maturation with respect to an external

DNA reference standard. The DNA-index shows a distinct decline of PCR amplificability during cheese ripening as consequence of a DNA loss of about 96%. DNA-index largely varied in the different types of cheeses: *e.g.* Mozzarella exhibited DNA indices of 1.6–9%, Feta 0.6–3.8%, matured hard cheese 1.1–10.4% and blue cheese 1.1–3.8%.

End Point PCR

Singleplex End Point PCR

A summary of methods, sequences tagged and results obtained with PCR in different dairy products, is presented in Table **2**.

Table 2. Summary of PCR methods for identification of milks employed in dairy products. sxPCR: single PCR; dxPCR: duplex PCR, txPCR: triplex PCR, qxPCR: quadruplex PCR, qsxPCR: quantitative single PCR, qdxPCR: quantitative duplex PCR, qtx: quantitative triplex PCR, qqxPCR: quantitative quadruplex PCR.

		End point PCR			
Samples	**Method**	**Primers**	**Species considered**	**Limit of detection (LOD)**	**References**
Milk	sxPCR-RFLP	β-casein	Sheep Goat Buffalo Cow	0.5% cow milk	[17]
Mozzarella and Fetacheeses.	sxPCR-RFPL	Mitochondrial Cyt*b*	Sheep Goat Buffalo Cow	Not reported	[22]
Mixture of cheeses	sxPCR	Mitochondrial D-loop	Cow Goat Sheep	0.1% cow milk	[23]
Milk	sxPCR	Mitochondrial Cyt*b*	Cow Goat	0.1% cow milk	[24]
Mozzarella cheeses	sxPCR	Mitochondrial Cyt*b*	Cow Buffalo	1.5% cow milk	[19]
Experimental Mozzarella cheeses with variable amounts of cow milk	sxPCR	Cytochrome oxidase subunit I	Cow Buffalo	0.5% cow milk	[26]
Binary mixes of goat and sheep milks	sxPCR	Mitochondrial 12S ribosomal gene	Goat Sheep	0.1% cow milk	[28]

(Table 2) contd.....

	End point PCR				
Samples	**Method**	**Primers**	**Species considered**	**Limit of detection (LOD)**	**References**
Cheeses with goat and sheep milk.	sxPCR	Mitochondrial 12S ribosomal gene	Goat Sheep	1% cow milk	[29]
Mozzarella cheese	dxPCR	Mitochondrial Cyt*b*	Cow Buffalo	1% cow milk	[32]
Experimental cow /goat cheeses 20 commercial goat cheeses	dxPCR	Mitochondrial 12S rRNA	Cow Goat	0.5% cow milk	[33]
19 cheeses from retails	txPCR	Mitochondrial 12S and 16S rRNA genes	Cow Sheep Goat	0.5% all milks considered	[34]
Binary milk mixtures Commercial dairy products (cheeses, yogurt, butter, milk, cream)	qxPCR	Mitochondrial sequences: Bos1: mt coordinates: 15255-15407 (A.N.* AF492351) Bos2: mt coordinates: 5117-5277 (A.N AF492351) Bubalus1: mt coordinates: 14314-14488 (A.N. NC_006295) Bubalus2: mt coordinates: 2989-3185 (A.N. NC_006295) Capra1: mt coordinates: 13935-14052 (A.N. GU295658) Capra2: mt coordinates: 7869-7995 (A.N. GU295658) Ovis1:: mt coordinates: 13943-14100 (A.N. AF010406) Ovis2:: mt coordinates: 9347-9503 (A.N. AF010406) CF (universal control): mt coordinates2911-3049 (A.N. AY526085)	Cow Sheep Goat Buffalo	1% all milks considered	[2]

(Table 2) contd.....

		End point PCR			
Samples	**Method**	**Primers**	**Species considered**	**Limit of detection (LOD)**	**References**
Experimental cow/sheep cheeses 10 commercial cow sheep/cheeses	dxPCR	Mitochondrial 12S and 16S rRNA genes	Cow Sheep	0.1% cow milk	[35]
Experimental cow/goat cheeses	dxPCR	Mitochondrial 12S and 16S rRNA genes	Cow Goat	0.1% cow milk	[36]
		Real time PCR			
Samples	**Method**	**Primers**	**Species considered**	**LOD/Limit of quantitation (LOQ)/Linear Dynamic range (LDR)**	**References**
Cow/ buffalo milk mixes Commercial cheeses	qdxPCR TaqMan based assay	Mitochondrial cyt*b*	Cow Buffalo	LOD 2% cow milk	[39]
Raw and pasteurised goat/sheep milks	qsxPCR TaqMan based assay	Mitochondrial 12S rRNAgenes	Goat Sheep	LOD: 0.6% goat milk LDR:0 .6-%10% goat milk	[41]
Raw and pasteurized cow/sheep milks	qsxPCR TaqMan based assay	Mitochondrial 12S rRNAgenes	Cow Sheep	LOD: 0.5% cow milk LDR:0.5-10% cow milk	[42]
Mozzarella cheese	qsxPCR SYBR Green/ TaqMan quantitative based assay	mitochondrial cyt*b* nuclear growth hormone (GH)	Cow Buffalo	LOD 0.1% cow milk LDR: 0.6-20% cow milk	[43]
Experimental cow/sheep/goat cheeses Commercial cheeses	qxPCR TaqMan based assay qtxPCR TaqMan based assay	Mitochondrial genes: tRNA Lys (cow); cyt*b* (sheep, goat, buffalo). Nulclear genes: β-actin gene (cow); prolactine receptor gene (sheep), specific insertion of a LINE-1 element in non coding region of GH (goat)	Cow Sheep Goat Buffalo	0.32% of DNA of each species 0.32-32% of DNA of each species	[16]

(Table 2) contd.....

| Samples | End point PCR | | | | |
	Method	Primers	Species considered	Limit of detection (LOD)	References
Liquido of governing of Mozzarella cheese	qsxPCR SYBR Green based assay	Mitochondrial COI	Cow Buffalo	LOD: 0.5% cow milk LDR: 0.5-30% cow milk	[44]
Binary milk mixes:cow/sheep, cow/goat, cow/buffalo Experimental cheeses made with binary milk mixture above. Commercial products (cheeses, yoghurt, cream, butter)	qxPCR SYBR Green based assay	Mitochondrial 12S rRNA Mitochondrial cyt*b*	Cow Sheep Goat Buffalo	LOD 0.1% cow milk LDR: milk: 0.5-10% to 1-25% LDR cheese: 0.1-5% to 1-10%	[45]
Commercial Feta cheeses	HRM	Bovine D-loop Caprine D-loop Sheep tRNA^{Lys}	Cow Sheep Goat	LOD 0.1% cow milk DR (Ct): 0.1-20% cow milk LDR (fluorescence): 1-50% goat milk	[49]

The first attempts to detect the animal DNA in dairy products were conducted tagging the gene for β-casein in DNA extracted from cow, sheep, goat and buffalo milk [17]. Digestion of PCR products with restriction enzymes showed polymorphism in the site of *Ava*II of amplicons derived from goat and cow DNA. Thus PCR, when combined with restriction fragment length polymorphism (RFLP), identified cow milk, in a mixture of goat and cow milk, with a limit of detection (LOD) of 0.5%. The same authors applied the single strand conformation polymorphism (SSCP) analysis on the milk mixtures and identified specific bands for sheep, cow, goat and buffalo DNA; however, this methodology was not adequate to detect the same levels of adulteration as PCR/RFLP.

Mozzarella cheese, which is produced in Southern Italy, is protected by the European legislation with the label PDO which says it must be made from water buffalo milk only [18]. Bovine milk or mixture of bovine and water buffalo milk are admitted but, in this case, Mozzarella is not protected by the PDO. The addition of undeclared bovine to buffalo milk is a common fraud, cow milk being less expensive [19].

Feta is a soft white cheese ripened in brine traditionally made in Greece. The European Commission [20] adopted Feta as a PDO, establishing that only cheeses produced in the mainland of Greece and the island of Lesbos, made with pure sheep milk or mixed with up to 30% of goat milk, may bear the name Feta. Similar brined cheeses produced in the eastern Mediterranean Sea, made partly or wholly with cow milk cannot be called Feta [21].

The milk composition of samples of Mozzarella and Feta, was assessed with PCR and primers designed on the mitochondrial gene for cytochrome b (cyt*b*), subsequent digestion of amplicons with restriction enzymes and identification of specific bands of DNA for each kind of milk [22].

These pioneering studies were followed by the identification of specific sequences of the four mammalian species usually used for milk production with a simplified procedure based on PCR only.

Actually it seems that mitochondrial (mt) DNA is more suitable than nuclear DNA in these kinds of analysis for at least two reasons: *i*) the higher variability of mtDNA provides a larger number of species-specific sequences; *ii*) the number of copies of mtDNA per cell is about 1000 times higher than those of nuclear DNA.

A commonly tagged mitochondrial sequence is the D-loop region that differs in cow, goat and sheep for several deletions. PCR conducted with primers designed on these different sequences, produced an amplicon specific of cow DNA, that allowed the detection of bovine milk in cheeses with a LOD of 0.1% [23].

Another mitochondrial target used frequently to detect the milk species in dairy products is the gene of cytochrome b (cyt*b*) (Table 2) that was firstly employed to detect contamination of cow milk in goat milk at very low percentage (0.1%) [24]. Amplification of cyt*b* was also employed to detect the addition of cow milk in Mozzarella cheese, that, as sentenced above, is one of the most common frauds in the dairy supply chain [19]. Laboratory tests on experimental cheeses demonstrated that PCR was capable of detecting down to 1.5% of cow milk in mixes with buffalo milk. This method applied to 30 commercial Mozzarella cheeses, collected in different areas of Southern Italy, revealed the presence of cow milk in 70% of samples, in agreement with other authors [25] and with the Central Inspectorate for Quality Protection and Repression of Frauds in alimentary products (ICQRF) of the Italian Ministry of Agriculture and Forestry (MiPAAF). A lower LOD of cow milk in Mozzarella was reported in experimental cheeses made with variable percentages of cow milk (0.5-30%) and in ten PDO cheeses purchased from different retailers using PCR primers designed on mitochondrial gene of cytochrome oxidase sub unit I (COI) [26]. The authors detected cow milk in the storage liquid of the experimental cheeses in a percentage of 0.5% and in one commercial Mozzarella.

A common forward (FW) and reverse (RV) species-specific primers designed on *cytb* were used to detect cow milk in mixes of goat and cow cheese [27]. In this case LOD was 1% of cow milk (Table **2**). The same method, applied to the analysis of 17 goats and 7 sheep commercial cheeses produced in Czech Republic, Slovakia, France, the Netherland and Italy, detected undeclared cow milk in three kinds of goat cheese and in one sheep cheese.

Though the discovery of undeclared cow milk in dairy product is the most relevant challenge, less common frauds were found: an example is the presence of goat milk, less expensive than sheep milk, in ovine cheeses [28]. Therefore, several efforts were also made to identify these two kinds of milk in dairy products. Targeting the mitochondrial 12S (mt12S) ribosomal gene, goat milk was detected with an LOD of 0.1% in fresh and heat treated mixtures of sheep and goat milk. With the same approach, the presence of caprine milk was assessed in sheep cheese; in this case, LOD was higher, 1%, as expected when analysing a more complex matrix [29].

Multiplex PCR

Multiplex PCR co-amplifies separate regions of DNA in a single reaction using two or more pairs of primer in the same reaction mix [30]. With this technique it is possible to detect more targets in a single reaction, but an optimisation of primer concentration is necessary to avoid any preferential amplification [31].

Pioneering attempts of duplex-PCR (dxPCR) were conducted to identify bovine and water buffalo DNA in Mozzarella cheese [32]. The authors targeted the gene *cytb*, using in a single PCR reaction one FV primer, common to cow and buffalo DNA and two RV primers specific for the two species. The dxPCR has produced two specific amplicons distinguishable on agarose gel. The method allowed the detection of cow milk in Mozzarella with LOD of 1%.

A dxPCR, targeting the mt12S gene was developed to detect cow milk in experimental goat cheeses prepared with different percentages of cow milk (0.1-40%) and in twenty commercial goat cheeses purchased in the Rio de Janeiro food marked [33]. With this method, the presence of cow milk was detected in experimental cheeses with LOD of 0.5% and in all commercial cheeses, but, since the assay was not quantitative, the authors were unable to determine whether it was a deliberate fraud or accidental contamination.

A triplex PCR (tx) with primers designed on mt12S and mt16S rRNA genes was employed to generate fragments of different lengths in goat, cow and sheep DNA

[34]. This assay, when applied to 19 cheeses obtained from the retailer's trade, showed that three of them contained cow milk not declared, while the composition of the others was in accordance with the label. Sensitivity of the method, tested on experimental samples of curd, made with fixed percentages of the above three milks was 0.5% for each species.

More recently, a quadruplexPCR (qxPCR) system able to detect simultaneously the four species of milk most commonly used in the dairy food chain has been developed [2]. The qxPCR analyses simultaneously targets different regions of the mitochondrial genome of cow, goat, sheep and buffalo, producing four amplicons that are separated by capillary electrophoresis instead of the agarose gel (Table 2). This method has a sensitivity of at least 1% of the relative amount of milk in binary mixtures produced with all combinations of milk species.

The qxPCR was used to test 96 dairy products in Portuguese food retail market, including PDO products. The survey showed that all PDO products were in accordance with the labelling but the 12.5% of the others it was not. The non-conformity consisted in the presence of cow milk in products that had to be made only from sheep milk (four cases) or in the absence of sheep milk in products that had to be made from cow and sheep milk (eight cases). Also cow and goat milk were detected in one sample described as containing exclusively sheep milk. These results support the hypothesis of a deliberate adulteration for economic benefits, being cow and goat milk less expensive than that of sheep.

Quantitative End Point PCR

The first attempts to use PCR to quantify the amount of different species of milk in dairy products were based on the linear relationship between the amount of DNA amplicon and the intensity of the bands on agarose gel. This relationship had been highlighted in a duplex end point PCR (dxPCR) to determine the amount of cow milk in experimental cheeses made with specific mixtures of cow and sheep milk [35, 36]. To overcome those variations which might have occurred during DNA preparation and affecting amplification, intensity of each band was normalized on the total intensity of both bands. A good correlation ($R^2 = 0.99$) was found between percentage of cow milk and relative band intensity; sensitivity was down to 0.1% of cow milk.

DxPCR, applied to ten commercial cheese has confirmed the information of the labels, with the exception of two samples which contained only cow milk, while, as stated, had to be made of cow and sheep milk. The application of the method to 17 commercialised goat cheeses, highlighted the fraudulent addition of cow milk (9–13%) in three samples labelled as pure goat and the omission of goat milk which was otherwise mentioned in the label.

Real Time PCR

Real time PCR is a powerful and highly accurate technology that allows the time course of the reaction to be followed during its occurrence, monitoring the emission of fluorescence when DNA is synthetized. The fluorescence may be produced either by probes that target specific gene or by intercalating dyes (*e.g.* SYBR Green I). Fluorescence becomes measurable when a sufficient number of DNA fragments are synthesized, it occurs at threshold cycle (Ct) or crossing point (Cp), the first cycle at which the instrument may discriminate the fluorescence produced by the reaction above its background.

Ct is inversely proportional to the number of template molecules within the sample: if this number is high, Ct will happen earlier, if it is low the Cp happens later (Fig. **2**).

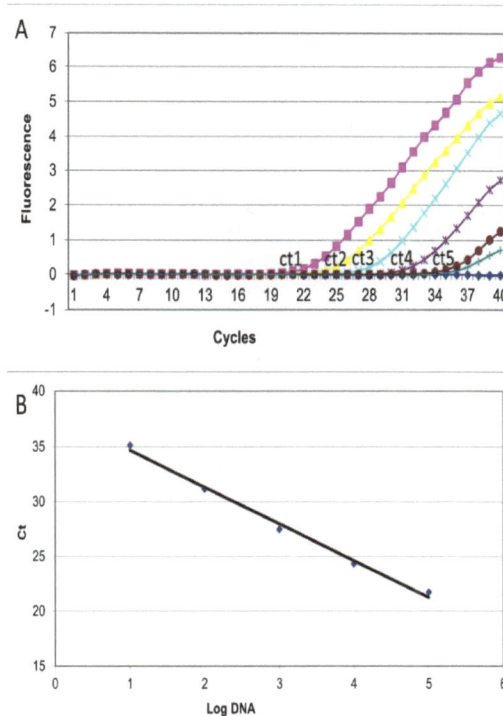

Fig. (2). A Plot of fluorescence in qPCR reaction with relative Ct. B. Reference curve correlating Ct with Log of known amount of DNA.

Relationship between Ct and the amount of template can be exploited to assess the number of DNA molecules present in a sample, building a reference curve in which known amounts of target DNA are plotted against the corresponding Ct.

From this curve it is possible to extrapolate the unknown amount of DNA present in a sample.

Real time PCR is commonly based on TaqMan fluorogenic probe [37] labelled with a reporter and a quencher dye which binds to a target DNA between two flanking primers. During PCR, the 5'–3' exonuclease activity of the Taq DNA polymerase cleaves the probe releasing the 5' reporter from the quencher; this results in an increase of fluorescence which is proportional to the amount of template DNA.

The use of fluorescence avoids the need for post-PCR processing, such as gel electrophoresis and DNA staining with ethidium bromide. Real time PCR can also be automated offering a large-scale sample processing.

Alternative to TaqMan probes, the intercalating dye SYBR Green (SG), is less expensive and more flexible. SG also allows for a confirmation of each amplicon specificity by melting curve analysis. Determination of temperature melting (Tm) of amplicon, which depends on the length and composition of nitrogen bases, is equivalent to detection of fragment size by gel electrophoresis. Real time PCR with SG has some disadvantages, as the formation of primer dimers that interfere with quantitation and its inhibitory effect on PCR at high concentrations [38] but offers the possibility of performing multiplex PCR if Tm of amplicons are different.

Qualitative Real Time PCR

A fluorescent qualitative test based on TaqManMGB (MinorGroove Binding) probes was developed to identify cow and buffalo DNA in a single reaction, exploiting a difference of only one base in the sequence of mtcyt*b* [39]. PCR was conducted with a common pair of primers for the two above species and two specific probes designed into the region carrying the mutation.

This method, initially applied to mixtures of cow and buffalo milk with a sensitivity of 2% for cow milk, was extended to cheeses purchased from retailers: forty commercial "Mozzarella di bufala campana" PDO, five "Mozzarella" made with cow milk, six samples allegedly made only with buffalo milk (no PDO) and one cheese made with both cow and buffalo milk. The analysis confirmed the label statement of the forty "Mozzarella di bufala campana" and of the five cow "Mozzarella". At variance, six labelled "pure buffalo" (no PDO) were not conform to the labelling: four samples showed the unmistakable presence of cow milk, and in the other two only cow milk was detected. In the cheese labelled as made with cow and buffalo milk only the cow milk was detected.

Quantitative Real Time PCR (qPCR)

Quantitation of DNA in food is a challenge because of the large fluctuations due to equipment, operators and variable amount and quality of DNA. Therefore, several replicates of the same assay, at the same (technical replicates) and different times (biological replicates) should be made. The variability among these replicates is defined by the coefficient of variation (CV), resulting from standard deviation/average value x 100. Of course, the lower is CV the higher is the robustness of the method. Following the guidelines of Codex Alimentarius (2004) [40], CV among technical replicates should be less than 25% and among biological replicates should be less than 33% for an assay to be acceptable.

Other important parameters of a qPCR assay are: *i*) the limit of quantitation (LOQ) that is the lowest quantity of target that can be quantitated with a CV < 25% among the technical replicates [39]; *ii)* the linear dynamic range (LDR) that is the interval in which the curve describing the correlation between Ct and amount of DNA is linear ($R^2 \geq 0.99$); *iii*) the efficiency that is the fold increase in PCR product per cycle; it is considered acceptable in a range between 90-110%.

An endogenous control that consists of a sequence of which is known a priori that is present in the samples in the same amount should be used to determine whether the different amounts of targets measured in different samples are real or if they are due to other factors such as the templates of different quality or quantity. The ratio between the amount of the specific target and the endogenous control allows the normalization of the data and therefore the reduction of the fluctuations due to the different sample preparation.

A qPCR with TaqMan probes was developed to quantitate goat milk in dairy products. The assay was based on the use of a probe targeting an endogenous sequence of mt12S gene, common to goat and sheep, and of species-specific primers designed on a sequence of the same gene, in this way the goat specific amplicon was normalized against the total DNA present in the sample [41]. This method allowed the quantitation of goat milk in raw and pasteurised milk mixtures, in a LDR from 0.6 to 10% with a CV among replicates less than 0.5%.

A similar approach, using the same gene sequences, was followed to quantitate cow milk in raw and heat-treated mixtures of cow and sheep milks: in this case the LDR was 0.5-10% of cow milk in the mixture [42].

The two qPCR methods, probes *vs.* SG, were compared to quantitate DNA of cow milk in Mozzarella cheeses [43]. The normalized procedure was based on the bovine mtcyt*b* to quantitate bovine DNA and on the nuclear growth hormone (GH) gene as universal endogenous control. LDR and efficiency of qPCR with

TaqMan Probes and SG were compared in DNA extracted from blood, milk and cheese. For blood and milk the LDR and efficiency of the two methods were substantially equivalent, but in cheese the use of probes significantly extended the LDR for both genes. For this reason, the TaqMan assay was preferred to quantitate cow milk in 64 commercials "Mozzarella di bufala" purchased at local supermarkets, dairy shops, or directly from producers. To conduct this analysis standard curve was constructed on DNA extracted from standard cheeses made with buffalo milk mixed with percentages of 0.1%, 0.6%, 1%, 2%, 5%, 10%, and 20% of cow milk. The LOD was 0.1% of cow milk, while the LDR was 0.6-20%.

The survey on commercial cheeses indicated that bovine milk was used, in variable amounts, in the majority of products, 79.7%, of the commercial samples, of which 76.5% were PDO. In 37.5% of the positive samples, presence of bovine milk was also quantitated [43].

Other authors tested the possibility of quantitate simultaneously cow, goat, sheep and buffalo DNA present in fresh and ripened cheese with a quadruplexPCR (qqxPCR) based on TaqMan probe [16]. The authors tested two methods: one based on the tagging cow, sheep, goat and buffalo mitochondrial sequences ("AllMilk") and the other on the tagging of single copy nuclear sequence of cow, sheep and goat ("AllCheese") (Table 2).

The authors evaluated also the influence of DNA calibrators used for building standard curve on accuracy of quantitation, testing serial dilutions of three of them: *i*) DNA extracted from meat (CDNA); *ii*) DNA extracted from fresh cheeses with known composition(CFC) and *iii*) DNA extracted from mature cheeses with known composition (CTS); these last two were defined "matrix-adapted calibrators". The performance of each method was evaluated in fresh and ripened experimental cheeses, made with known proportions of different milks, in a ring test involving eleven laboratories. By using the "AllCheese" method with the CDNA calibrator, the bovine milk proportions were significantly under-estimated; this was explained by the low number of somatic cells in cow milk in comparison to the sheep and goat milk, which brings to preferential amplification of the DNA of these species in qqxPCR. This effect is reduced using matrix-specific calibrators CFC and CTS. Differently, quantitation of the other milks was not significantly affected by the calibrators.

At variance from "AllCheese", the method "AllMilk" gave an over-estimation of the cow milk content, but the quantitation of all milks approached the real values using matrix-specific calibrators, CTS and CFC instead of CDNA.

"AllMilk" method was used for detection and quantitation of cow, sheep, goat and buffalo milk in thirty-four cheeses purchased at market. Calculations was performed using both CDNA and matrix adapted calibrators. The two methods gave the same results: only seven cheeses had a composition consistent with the label: the discrepancies were due mostly to undeclared presence of cow milk, even at percentages < 1%. However, it has been observed that the relative amounts of goat and water buffalo milks varied significantly according to the calibration mode. Unfortunately, the true content of milks in these cheeses are unknown, and the accuracy of the two calibration approaches could not be estimated.

Instead of TaqMan probe assay, qPCR based on SG was used by other authors to detect and quantitate cow milk in Mozzarella cheese, using primers targeting a specific sequence in COI gene [44]. DNA was extracted from storage liquid of cheeses containing increasing percentages of bovine milk, from 0.5 to 30%. The lowest amount of bovine DNA detected in the cheese, corresponded to samples containing 0.5% of cow milk.

The possibility to quantitate the four milk species in dairy products using qqxPCR based on SG was explored [45]. The method was firstly tested in binary mixes of cow milk combined with each milk of the other species in percentages varying from 0.1 to 25% and on cheeses made with the same mixes. Specific primers for the four animal species were designed on sequences of 12S and cyt*b* genes and combined in suitable concentrations to detect the kind of milk in mixes and cheeses. The qqxPCR allowed the detection of cow milk in a percentage of 0.1%.

Applicability of qqxPCR platform was tested on commercial dairy products made with different milk composition, comparing the results obtained with singleplex PCR (qsxPCR) in which the four primer pairs were utilized separately.

Out of 26 dairy commercial products analysed with the qqxPCR, 21 (80%) gave similar profiles to those obtained with qsxPCR. The 61% of products analysed with qqxPCR showed a composition of milks conform to the label; this percentage raised to 77%, when the samples were analysed with qsxPCR. The other products resulted only partially consistent to the label: in the majority of cases the analysis showed the presence of undeclared cow milk, in others only one kind of milk (sheep or cow) or a number of milks lower than declared was detected.

Differently from multiplex qPCR based on TaqMan probes, multiplex qPCR based on SG was not used for quantitation. However, the possibility of quantitating cow milk with qqxPCR by establishing a correlation between the peak areas under the dissociation curves of each amplicon and the percentages of cow milk was explored [45]. For milk mixes the correlation values (R^2) were rather satisfactory, being > 0.9 in ranges varying from 0.5-10% to 1-25% (Fig. **3**) in three experiments. In cheeses, in only one experiment R^2 was > 0.9 (Fig. **3**), while in the others it was <0.7; with cheese the range of quantitation varied from 0.1-5% (cow/goat) to 1-10% (cow/sheep and cow/buffalo). These results showed that the procedure was feasible with a standard curve made with reference mixes of milk at fixed percentages, while quantitation in cheeses is complicated by the higher complexity of the matrix [16]. The availability of appropriate standards for dairy products, some available in EU [9], may help to estimate whether a milk compositional labelling of PDO cheese is feasible or not.

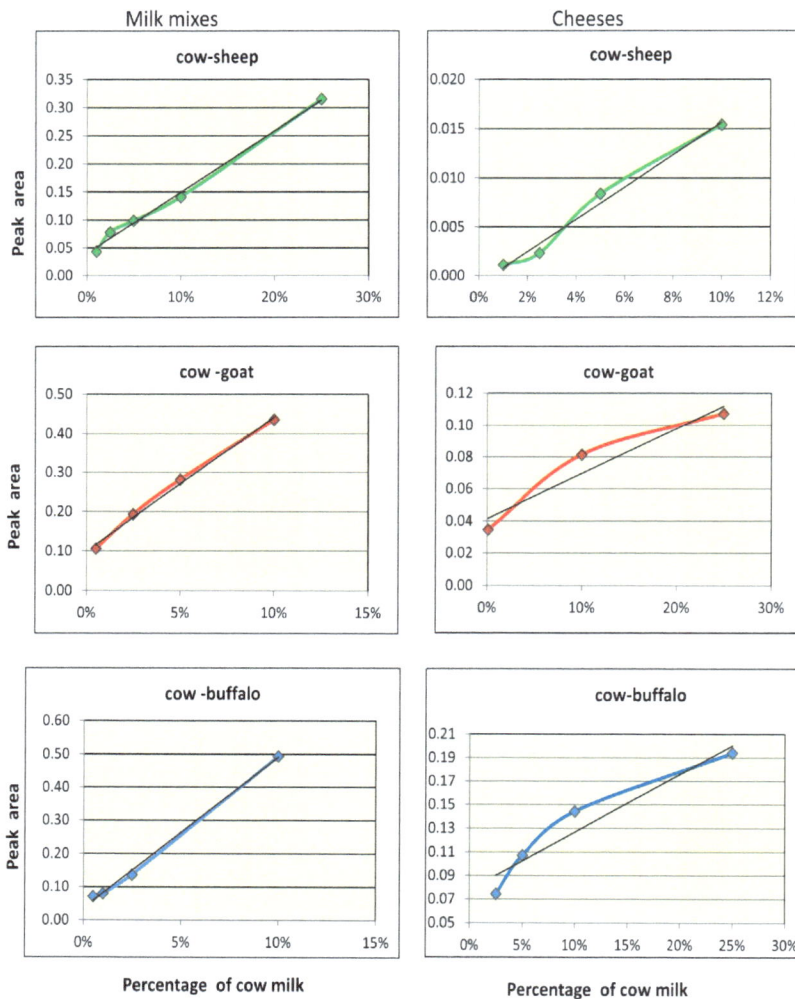

Fig. (3). Regression curves correlating the percentages of cow milk, in milk mixes and in cheeses, with the areas of peaks of derivative of amplicon dissociation curves. Reprinted from Agrimonti *et al.* [45].

High resolution melting (HRM) analysis [46] is a DNA based method that allows genotyping and fingerprinting by discriminating DNA sequence variants such as single nucleotide polymorphisms (SNPs) and small insertion and deletions (INDEL). HRM differs from standard SG dye melt curve analysis in three ways: *i*) saturating and brighter double-stranded DNA binding dyes, in particular SYTO®9, are used; *ii*) instruments performing HRM collect more data point than in a standard melting curve; *iii*) software uses new fluorescence normalization algorithms and plots [47, 48].

HRM with specific mitochondrial primers (Table **2**) was conducted for the rapid detection and quantitation of bovine, ovine and caprine milk in some samples of Feta [49], the Greek cheese described in the section "Singleplex end point PCR".

Cheese samples including eight commercial Feta PDO, three samples of sheep–goat's cheese (no PDO) and one sample allegedly made only with goat milk (no PDO) were analysed by HRM, using specific cow, sheep and goat primers. The assay, preliminary tested on milk mixtures containing goat and sheep milk added with fixed percentages of cow milk, gave identification and accurate quantitation of bovine milk down to 0.1% in milk mixtures, with the LDR between 0.1-20%. The method, when applied to commercial cheeses, showed that only one sample contained undeclared cow milk which was considered an accidental contamination.

In Feta cheese the presence of goat milk should be detected and quantitated to ensure that does not exceed 30% permitted by the legislation. At this purpose, authors above [49] explored the possibility of quantitating DNA species measuring the fluorescence intensity at inflection points of dissociation curve instead of Ct. This method was therefore tested on reference cheeses containing percentages of goat milk, ranging from 1 to 50%, in laboratory made cheeses. A good correlation, $R^2 = 0.945$, was found between fluorescence intensity at inflection point and goat milk percentage in samples, confirming that fluorescence intensities, as well as the peak area under dissociation curve may be used for quantitation of DNA present in dairy products.

CONCLUDING REMARKS

Use of DNA analysis to find adulterations in dairy products was not possible before the development of PCR. In fact, the usual methods based on nucleic acid hybridisation, like Southern blotting, required high amounts of DNA of good quality difficult to obtain from dairy products. After development of PCR the possibility of identifying plant and animal species in food, targeting residual DNA, originated the "Food genomics" [50, 51]. In the dairy field the first attempts of "Food genomics" were conducted in the 1997 [17] by PCR combined with RFLP or SSCP analysis. From there on, real time PCR fastened the analysis, avoiding agarose gel electrophoresis and offering also the possibility of quantitating target DNA.

Many papers reported application of PCR, end point and real time targeting mitochondrial genes to detect very low traces of DNA corresponding to 0.1-0.5% of cow milk in dairy products, consistently with EU official method [7] based on isoelectric focusing of γ-casein (LOD required $\leq$ 0.5%). The capacity with multiplex PCR to simultaneously detect DNA of the four mammalian species employed in dairy production is also a consolidated acquisition.

The possibility of quantitating the presence of less expensive milks like cow or goat in PDO cheeses, like Feta or Mozzarella, in which the mentioned milks, cannot be above a certain threshold, is still controversial. Several papers reported the possibility of quantitating the kind of milks in milk mixtures or in reference cheeses with known percentages of milks, but some problems remain to be solved for commercial products, because there are seasonal variations in the number of residual mammalian cells and/or because of the quality of DNA extracted, especially poor in ripened cheeses. So far the method more employed for the quantitation of mammalian DNA is qPCR based on TaqMan chemistry, that allows also for simultaneous quantitation of two or more targets, but the possibility of reaching a satisfactory quantitation with multiplex real time based on less expensive SG or HRM remains an open option.

CONSENT FOR PUBLICATION

Not applicable.

CONFLICT OF INTEREST

The author (editor) declares no conflict of interest, financial or otherwise.

ACKNOWLEDGEMENT

This publication was financially supported by the European Commission in the Communities 6th Framework Programme, Project TRACEBACK (FOOD-CT-036300). The authors also wish to acknowledge contribution of Emilia-Romagna Region (POR FSE 2007-2013).

REFERENCES

[1]   Directive 2000/13/EC of the European Parliament and of the Council of 20 March 2000 on the approximation of the laws of the Member States relating to the labelling, presentation and advertising of foodstuffs. Official Journal of the European Communities 2000; 29-42.

[2]   Gonçalves J, Pereira F, Amorim A, van Asch B. New method for the simultaneous identification of

cow, sheep, goat, and water buffalo in dairy products by analysis of short species-specific mitochondrial DNA targets. J Agric Food Chem 2012; 60(42): 10480-5.
[http://dx.doi.org/10.1021/jf3029896] [PMID: 23025240]

[3]   Sampson HA. 9. Food allergy. J Allergy Clin Immunol 2003; 111(2) (Suppl.): S540-7.
[http://dx.doi.org/10.1067/mai.2003.134] [PMID: 12592300]

[4]   Fiocchi A, Dahdah L, Albarini M, Martelli A. Cow's milk allergy in children and adults. Chem Immunol Allergy 2015; 101: 114-23.
[PMID: 26022871]

[5]   Council Regulation (EEC) No 2081/92 on the protection of geographical indications and designations of origin for agricultural products and foodstuffs. Official Journal of the European Union, L 208, 1992, pp. 0001-0008.

[6]   Commission Regulation (EU) No 1151/2012 of the European Parliament and of the Council of 21 November 2012 on quality schemes for agricultural products and foodstuffs. Off J Eur Union L 2012; 349: 1-29.

[7]   Commission Regulation (EC) No 273/2008. Reference method for the detection of cows' milk and caseinate in cheeses from ewes' milk, goats' milk or buffalos' milk or mixtures of ewes', goats' and buffalos' milk. Official Journalof the European Union 2008; L88: 53-61.

[8]   Molina E, Martín-Álvarez JP, Ramos M. Analysis of cows', ewes' and goats' milk mixtures by capillary electrophoresis: quantitation by multivariate regression analysis. Int Dairy J 1999; 9: 99-105.
[http://dx.doi.org/10.1016/S0958-6946(99)00028-X]

[9]   Mayer HK, Bürger J, Kaar N. Quantification of cow's milk percentage in dairy products - a myth? Anal Bioanal Chem 2012; 403(10): 3031-40.
[http://dx.doi.org/10.1007/s00216-012-5805-1] [PMID: 22349339]

[10]   Hurley IP, Ireland HE, Coleman RC, Williams JHH. Application of immunological methods for the detection of species adulteration in dairy products. Int J Food Microbiol 2004; 39: 873-8.

[11]   Ferreira IM, Caçote H. Detection and quantification of bovine, ovine and caprine milk percentages in protected denomination of origin cheeses by reversed-phase high-performance liquid chromatography of beta-lactoglobulins. J Chromatogr A 2003; 1015(1-2): 111-8.
[http://dx.doi.org/10.1016/S0021-9673(03)01261-5] [PMID: 14570324]

[12]   Cozzolino R, Passalacqua S, Salemi S, Garozzo D. Identification of adulteration in water buffalo mozzarella and in ewe cheese by using whey proteins as biomarkers and matrix-assisted laser desorption/ionization mass spectrometry. J Mass Spectrom 2002; 37(9): 985-91.
[http://dx.doi.org/10.1002/jms.358] [PMID: 12271441]

[13]   de la Fuente MA, Juárez M. Authenticity assessment of dairy products. Crit Rev Food Sci Nutr 2005; 45(7-8): 563-85.
[http://dx.doi.org/10.1080/10408690490478127] [PMID: 16371328]

[14]   Rossen L, Nørskov P, Holmstrøm K, Rasmussen OF. Inhibition of PCR by components of food samples, microbial diagnostic assays and DNA-extraction solutions. Int J Food Microbiol 1992; 17(1): 37-45.
[http://dx.doi.org/10.1016/0168-1605(92)90017-W] [PMID: 1476866]

[15]   Pirondini A, Bonas U, Maestri E, Visioli G, Marmiroli M, Marmiroli N. Yield and amplificability of different DNA extraction procedures for traceability in the dairy food chain. Food Control 2010; 21(5): 663-8.
[http://dx.doi.org/10.1016/j.foodcont.2009.10.004]

[16]   Rentsch J, Weibel S, Ruf J, Eugster A, Beck K, Köppel R. Interlaboratory validation of two multiplex quantitative real-time PCR methods to determine species DNA of cow, sheep and goat as a measure of milk proportions in cheese. Eur Food Res Technol 2013; 236: 217-27.
[http://dx.doi.org/10.1007/s00217-012-1880-y]

[17] Plath A, Krause I, Einspanier R. Species identification in dairy products by three different DNA-based techniques. Z LebensmUntersForsch A 1997; 205: 437-41.

[18] Commission Regulation (EC) No 2527/98 supplementing the Annex to Regulation (EC) No 2301/97 on the entry of certain names in the 'Register of certificates of specific character' provided for in Council Regulation (EEC) No 2082/92 on certificates of specific character for agricultural products and foodstuffs (Text with EEA relevance). Official Journal of the European Union L317, 1998, pp. 0014 – 0018.

[19] Di Pinto A, Conversano MC, Forte VT, Novello L, Tantillo GM. Detection of cow milk in buffalo "mozzarella" by polymerase chain reaction (PCR) assay. J Food Qual 2004; 27: 428-35.
[http://dx.doi.org/10.1111/j.1745-4557.2004.00662.x]

[20] Commission Regulation (EC) No. 1829/2002 amending the Annex to Regulation (EC) No. 1107/96 with regard to the name 'Feta'. Off J Eur Union L 2002; 227: 10-4.

[21] Mauropoulos AA, Arvanitoyannis IS. Implementation of hazard analysis critical control point to Feta and Manouri cheese production lines. Food Control 1999; 10: 213-9.
[http://dx.doi.org/10.1016/S0956-7135(99)00021-3]

[22] Branciari R, Nijman IJ, Plas ME, Di Antonio E, Lenstra JA. Species origin of milk in Italian mozzarella and Greek feta cheese. J Food Prot 2000; 63(3): 408-11.
[http://dx.doi.org/10.4315/0362-028X-63.3.408] [PMID: 10716574]

[23] Maudet C, Taberlet P. Detection of cows' milk in goats' cheeses inferred from mitochondrial DNA polymorphism. J Dairy Res 2001; 68(2): 229-35.
[http://dx.doi.org/10.1017/S0022029901004794] [PMID: 11504387]

[24] Bania J, Ugorski M, Polanowski A, Adamczyk E. Application of polymerase chain reaction for detection of goats' milk adulteration by milk of cow. J Dairy Res 2001; 68(2): 333-6.
[http://dx.doi.org/10.1017/S0022029901004708] [PMID: 11504396]

[25] Dalvit C, De Marchi M, Cassandro M. Genetic traceability of livestock products: A review. Meat Sci 2007; 77(4): 437-49.
[http://dx.doi.org/10.1016/j.meatsci.2007.05.027] [PMID: 22061927]

[26] Feligini M, Bonizzi I, Curik CV, Parma P, Greppi GF, Enne G. Detection of adulteration in Italian cheese using mitochondrial DNA templates as biomarkers. Food Technol Biotechnol 2005; 43: 91-5.

[27] Maskova E, Paulickova I. PCR-based detection of cow's milk in goat and sheep cheeses marketed in the Czech Republic. Czech J Food Sci 2006; 24: 127-32.
[http://dx.doi.org/10.17221/3307-CJFS]

[28] López-Calleja I, González I, Fajardo V, *et al.* Application of polymerase chain reaction to detect adulteration of sheep's milk with goats' milk. J Dairy Sci 2005; 88(9): 3115-20.
[http://dx.doi.org/10.3168/jds.S0022-0302(05)72993-3] [PMID: 16107400]

[29] Lopez-Calleja I, Diaz I, Gonzalez A, *et al.* Application of a polymerase chain reaction to detect adulteration of ovine cheeses with caprine milk. Eur Food Res Technol 2007; 225: 345-9.
[http://dx.doi.org/10.1007/s00217-006-0421-y]

[30] Chamberlain JS, Gibbs RA, Ranier JE, Nguyen PN, Caskey CT. Deletion screening of the Duchenne muscular dystrophy locus *via* multiplex DNA amplification. Nucleic Acids Res 1988; 16(23): 11141-56.
[http://dx.doi.org/10.1093/nar/16.23.11141] [PMID: 3205741]

[31] Elnifro EM, Ashshi AM, Cooper RJ, Klapper PE, Multiplex PCR. Multiplex PCR: optimization and application in diagnostic virology. Clin Microbiol Rev 2000; 13(4): 559-70.
[http://dx.doi.org/10.1128/CMR.13.4.559-570.2000] [PMID: 11023957]

[32] Rea S, Chikuni K, Branciari R, Sangamayya RS, Ranucci D, Avellini P. Use of duplex polymerase chain reaction (duplex-PCR) technique to identify bovine and water buffalo milk used in making

mozzarella cheese. J Dairy Res 2001; 68(4): 689-98.
[http://dx.doi.org/10.1017/S0022029901005106] [PMID: 11928964]

[33]   Golinelli LP, Carvalho AC, Casaes RS, *et al.* Sensory analysis and species-specific PCR detect bovine milk adulteration of frescal (fresh) goat cheese. J Dairy Sci 2014; 97(11): 6693-9.
[http://dx.doi.org/10.3168/jds.2014-7990] [PMID: 25200782]

[34]   Bottero MT, Civera T, Nucera D, Rosati S, Sacchi P, Turi RM. A multiplex polymerase chain reaction for identification of cow', goat' and sheep's milk in dairy products. Int Dairy J 2003; 13: 277-82.
[http://dx.doi.org/10.1016/S0958-6946(02)00170-X]

[35]   Mafra I, Ferreira IM, Faria MA, Oliveira BP. A novel approach to the quantification of bovine milk in ovine cheeses using a duplex polymerase chain reaction method. J Agric Food Chem 2004; 52(16): 4943-7.
[http://dx.doi.org/10.1021/jf049635y] [PMID: 15291455]

[36]   Mafra I, Roxo A, Ferreira I, Oliveira B. A duplex polymerase chain reaction for the quantitative detection of cows' milk in goats' milk cheese. Int Dairy J 2007; 17: 1132-8.
[http://dx.doi.org/10.1016/j.idairyj.2007.01.009]

[37]   Holland PM, Abramson RD, Watson R, Gelfand DH. Detection of specific polymerase chain reaction product by utilizing the 5'----3' exonuclease activity of *Thermus aquaticus* DNA polymerase. Proc Natl Acad Sci USA 1991; 88(16): 7276-80.
[http://dx.doi.org/10.1073/pnas.88.16.7276] [PMID: 1871133]

[38]   Giglio S, Monis P, Saint C. Demonstration of preferential binding of SYBR Green I to specific DNA fragments in real-time multiplex PCR. Nucleic Acids Res, 2003, 31, e31 (2-5).
[http://dx.doi.org/10.1093/nar/gng135]

[39]   Dalmasso A, Civera T, La Neve F, Bottero T. Simultaneous detection of cow and buffalo milk in mozzarella cheese by real-time PCR assay. Food Chem 2010; 124: 362-6.
[http://dx.doi.org/10.1016/j.foodchem.2010.06.017]

[40]   Codex AlimentariusCommision Criteria for the method for the detection and identification of foods derived from biotechnology General approach and criteria for the methods CX/MAS 04/10. Rome: Food and Agriculture Organization 2004.

[41]   Lopez-Calleja I, Gonzales I, Fajardo J, *et al.* Quantitative detection of goats' milk in sheep's milk by real-time PCR. Food Control 2007; 18: 1466-73.
[http://dx.doi.org/10.1016/j.foodcont.2006.11.006]

[42]   Lopez-Calleja I, Gonzales I, Fajardo J, *et al.* Real-time TaqMan PCR for quantitative detection of cow' milk in ewes' milk mixtures. Int Dairy J 2007; 17: 729-36.
[http://dx.doi.org/10.1016/j.idairyj.2006.09.005]

[43]   Lopparelli RM, Cardazzo B, Balzan S, Giaccone V, Novelli E. Real-time TaqMan polymerase chain reaction detection and quantification of cow DNA in pure water buffalo mozzarella cheese: method validation and its application on commercial samples. J Agric Food Chem 2007; 55(9): 3429-34.
[http://dx.doi.org/10.1021/jf0637271] [PMID: 17419643]

[44]   Feligini M, Alim N, Bonizzi I, Enne G, Alesandri R. Detection of cow milk in water buffalo cheese by Sybr green realtimepcr: sensitive test on governing liquid samples. Pak J Nutr 2007; 6: 94-8.
[http://dx.doi.org/10.3923/pjn.2007.94.98]

[45]   Agrimonti C, Pirondini A, Marmiroli M, Marmiroli N. A quadruplex PCR (qxPCR) assay for adulteration in dairy products. Food Chem 2015; 187: 58-64.
[http://dx.doi.org/10.1016/j.foodchem.2015.04.017] [PMID: 25976998]

[46]   Wittwer CT, Reed GH, Gundry CN, Vandersteen JG, Pryor RJ. High-resolution genotyping by amplicon melting analysis using LCGreen. Clin Chem 2003; 49(6 Pt 1): 853-60.
[http://dx.doi.org/10.1373/49.6.853] [PMID: 12765979]

[47]   Graham R, Liew M, Meadows C, Lyon E, Wittwer CT. Distinguishing different DNA heterozygotes

by high-resolution melting. Clin Chem 2005; 51(7): 1295-8.
[http://dx.doi.org/10.1373/clinchem.2005.051516] [PMID: 15905310]

[48]   Reed GH, Kent JO, Wittwer CT. High-resolution DNA melting analysis for simple and efficient molecular diagnostics. Pharmacogenomics 2007; 8(6): 597-608.
[http://dx.doi.org/10.2217/14622416.8.6.597] [PMID: 17559349]

[49]   Ganopoulos I, Sakaridis I, Argiriou A, Madesis P, Tsaftaris A. A novel closed-tube method based on high resolution melting (HRM) analysis for authenticity testing and quantitative detection in Greek PDO Feta cheese. Food Chem 2013; 141(2): 835-40.
[http://dx.doi.org/10.1016/j.foodchem.2013.02.130] [PMID: 23790855]

[50]   Marmiroli N, Peano C, Maestri E. Advanced PCR techniques in identifying food components. 2003.
[http://dx.doi.org/10.1533/9781855737181.1.3]

[51]   Agrimonti C, Marmiroli N. DNA extraction and analysis from feeds and foods: a tool for traceability.DNA binding and DNA extraction. Nova Science Publishers, Inc. 2011; pp. 181-204.

CHAPTER 3

Inspection of Colorant Adulteration by Modern LC Mass Spectrometry

Mingchih Fang, Chia-Fen Tsai and **Hwei-Fang Cheng**[*]

Division of Research and Analysis, Taiwan Food and Drug Administration, Taipei City, Taiwan

Abstract: This chapter introduces modern mass spectrometry for the detection of dyes in foods. An LC/MS/MS method for the determination of 20 synthetic dyes is first described. Followed high resolution mass spectrometry (HRMS) demonstrates the modern screening method for illegal dye adulteration. There is an example of semi-targeted/non-targeted detection and structure confirmation of a non-permitted dye, diethyl yellow, by quadrupole-orbitrap HRMS. Mass spectrometry operated on selected reaction monitoring (SRM) is gold standard normally used for the detection of analytes such as pesticides, veterinary medicines and additives such as dyes in complex matrix. However, SRM only allows signals which fit target list to be recorded. It will lose all information of semi-target, non-target or even unknown. A new concept called data-independent acquisition (DIA) overcomes the limitations of SRM and increases the possibility of semi-target and non-target detection. This chapter includes the application of DIA for simultaneously screening and confirmation of dyes in various foods.

Keywords: Adulteration, Carminic acid, Colorants, Data-independent acquisition, Dye, Non-targeted, Orbitrap, Semi-targeted, Sudan dyes, Sulforhodamine B.

INTRODUCTION

Food appearance, especially color, along with flavor and taste strongly influences satiety. Food needs its right color. Unappropriated color of food leads to adverse effects on consumption and digestion. People aware and report the weird taste of green ketchup provided by instant food chain on St Patrick's Day, even if the green ketchup is formulated by the same ingredient with red ketchup but additional green dye is added. However, other people especially kids think the colored green ketchup is fun and exciting.

Natural colors/dyes such as anthocyanins, carotrnoids, betalaines, and chloro-

[*] **Corresponding author Hwei-Fang Cheng:** Division of Research and Analysis, Taiwan Food and Drug Administration, Taipei City, Taiwan, 161-2, Kunyang St., Taipei City 11561, Taiwan; Tel: 886-227877002; E-mail: rmhfcheng@fda.gov.tw

phylls are unstable, hence synthetic dyes are added to give the expected color and assure the uniformity from batch to batch during food processing. Coloring food has been recognized as an economical means of restoring quality and increasing price based on the misleading of outward appearance and enhancing of characteristically natural color. It is possible that synthetic food dyes added into foods exceed the authorized scopes and levels. Monitoring of the synthetic dyes becomes therefore particular important as they can contribute to carcinogens and allergens [1, 2].

Various analytical methods have been proposed for the determination of dyes in food stuffs. High-performance liquid chromatography (HPLC) with ultraviolet detector [3], diode array detector [4, 5], and mass spectrometry [6, 7] are the most common methods. Others such as enzyme-linked immunosorbent assay (ELISA) [8] and nuclear magnetic resonance spectrometry (NMR) with multivariate classification [9] have also been applied.

This chapter aims to introduce modern analytical methods for the detection of dyes in foods. A basic LC/MS/MS method for the determination of 20 synthetic dyes is first described. Followed simultaneous detection of near 40 dyes by HPLC coupled with high resolution mass spectrometry (HRMS) demonstrates the modern screening method for mostly illegal dyes in foodstuffs. Six dyes including Carmine, Sunset yellow (E110), Srocein orange G, Orange II, Sulforhodamine B, and Amaranth are discussed. In addition, non-targeted detection of Diethyl yellow by quadrupole-Orbitrap HRMS is described. Finally, a newly developed data-independent acquisition (DIA) concept overcomes the limitations of data-dependent acquisition (DDA) such as selected reaction monitoring (SRM) also known as multiple reaction monitoring (MRM) is applied into dyes analysis. DIA employs all ion fragmentation (AIF) generated multiplexed data sets that allowed to obtain information for dye identification while SRM-like ion chromatographs can still be subsequently extracted or reconstructed from the raw data. The advantages and limitations are being considered.

DETECTION OF 20 SYNTHETIC DYES BY HPLC-MS/MS

International Agency for Research on Cancer has confirmed the toxicities of synthetic azo dyes, such as Sudan I - IV [10] as category 3 carcinogen to humans. Although Sudan dyes have been forbidden in recent decades, it can still be found in various food products in many countries [11]. Followed HPLC-MS/MS method describes a single step extraction protocol offered simple and rapid sample preparation for the determination of 20 dyes in chili powders and syrup-preserved fruits.

HPLC-MS/MS DETERMINATION

The determination of twenty dyes was performed on an Acclaim® Polar Advantage C16 (3 μm, 120Å, 4.6 ×150 mm) column connected to an UltiMate® 3000 Standard LC System (Thermo Fisher Scientific Inc., MA, USA) coupled with a triple quadrupole mass spectrometer API 3200 (AB SCIEX, MA, USA). Eleven dyes including Dimethyl Yellow, Fast Garnet GBC, Para Red, Sudan I-IV, Sudan Orange G, Sudan Red 7B, Sudan Red B, and Sudan Red G were detected by positive electrospray ionization (ESI+), and other nine dyes including New Coccine, Indigo Carmine, Erythrosine, Tartrazine, Sunset Yellow FCF, Fast Green FCF, Brilliant blue FCF, Allura Red AC, and Amaranth were detected by negative ion electrospray ionization (ESI-). The mobile phase was consisted of acetonitrile (A) and 20 mM ammonia acetate buffer with 1% acetic acid (B). Linear gradient elution was programmed as following: 0-3 min, 98-50% B; 3-8 min, 50-0% B; 8-15 min, 0% B; 15-15.5 min, 0-98% B; 15.5-20 min, 98% B. Flow rate was set at 1 mL/min. Standard solutions of individual dyes was prepared in dimethyl sulfoxide DMSO (1 mg/mL) and further diluted with methanol to obtain working solution of 10 μg/mL. Samples, chili powders and syrup-preserved fruits, 10g was extracted with 50 mL acetonitrile twice. The combined extract was centrifuged for 5 min at 15°C. The supernatant was filtered through a 0.45 μm PVDF membrane filter prior to analysis.

Target dyes were identified by their retention times (RTs) and selected ions in selected reaction monitoring (SRM) mode. The instrumental settings and the limit of quantitation were summarized in Table **1**. This method offered rapid sample preparation and HPLC analysis. The 20 min LC-MS/MS method under SRM mode was able to detect all target compounds in a single run with LOQ between 0.001 and 1 mg/kg. In overall, the method was suitable for routine dyes testing and surveillance programs for the control of the presence of dyes in chili powders and syrup-preserved fruits. Amaranth (E123) which was an illegal dye in some countries such as United States was detected in a syrup-preserved fruit sample in this survey.

Table 1. Values of the instrumental settings, limit of quantitation (LOD) of each compound.

Compound	RT (min)	Precursor ion (m/z)	Product ion (m/z)	Collision energy (V)	LOD (mg kg^{-1})
Sudan Orange G	8.10	215	122*, 93	31, 21	0.01
Dimethyl Yellow	9.57	226	120*, 105	43, 25	0.005
Sudan I	10.25	249	156*, 93	33, 21	0.01
Sudan II	11.18	277	121*, 106	25, 55	0.005

(Table 1) contd.....

Compound	RT (min)	Precursor ion (m/z)	Product ion (m/z)	Collision energy (V)	LOD (mg kg⁻¹)
Sudan Red G	10.08	279	123*, 108	23, 47	0.001
Para Red	9.94	294	156*, 128	21, 35	0.01
Sudan III	10.15	353	197*, 128	27, 51	0.025
Sudan Red 7B	14.55	380	183*, 115	21, 65	0.001
Sudan IV	10.83	381	143*, 104	37, 83	0.05
Sudan Red B	10.81	381	156*, 134	27, 27	0.02
Fast Grarnet GBC	9.08	226	107*, 91	29, 31	0.01
Fast Green FCF	4.63	381	170*, 80	-98, -38	1
Tartrazine	2.22	244	198*, 80	-62, -20	0.5
New Coccine	4.18	268	206*, 80	-40, -18	1
Indigo Carmine	2.88	226	105*,198	-27, -53	0.25
Brilliant Blue	4.67	373	170*, 80	-92, -42	0.5
Sunset Yellow	4.25	203	207*, 171	-20, -20	0.25
Erythrosine	7.26	834	227*, 127	-91, -84	1
Allura Red AC	4.41	225	136*, 80	-59, -34	1
Amaranth	3.86	279	221*, 206	-30, -22	1

*qualifier ions.

SEMI-TARGETED/NON-TARGETED DETECTION OF DYES BY HPLC/HRMS, A SCREENING METHOD

The analysis of contaminants in food sample is a very difficult task due to its complicate matrix. Traditional food trace analysis focus on a single compound or a group of selected targets, developing and optimizing suitable method for each step such as, sample extraction, cleaning, concentration and instrument analysis. This analytical mode offers high sensitivity and reliability, thus most of official method and standard method are presented in this way in recent decades. However, the traditional trace analysis is facing a major drawback as it will miss compounds which were not selected in list at the beginning, thus, non-target compounds or some harmful contaminants even present in food at high levels will not able to be detected. Food preparation involves know-how and secret recipe. It is possible that in certain foods the concentrations of known compounds are not high enough to explain some of the sensory effects such as flavor, aroma and color. Therefore, non-target screening methods are very important tools for the determination of non-target compounds in foods. High resolution mass spectrometry (HRMS), which enables accurate mass measurement at high resolving power, is rapidly developed and more affordable in last decade. The main HRMS techniques today include time-of flight mass spectrometry (TOF) and Orbitrap Fourier-transform mass spectrometry (Orbitrap MS). The obtained accurate mass of a compound sufficiently elucidates its elemental composition and refers to an unknown compound identification.

We developed a new method for screening of 10 sweetens 12 preservatives and 40 dyes in food samples utilizing high resolution Orbitrap spectrometry. Similar work which simultaneously detected 68 dyes in wines was reported by Jia and his co-works [11]. A detection list (Table **2**) also called semi-targets was established for retrospective data analysis in our study. Instrument was operated at full-scan mode to obtain the accurate mass information of all possible analytes which also allowed for semi-targeted/non-targeted retrospective detection.

Instruments and Separation

The separation was performed on an UHPLC-ESI-Orbitrap-MS system consisted of an Accela 1250 LC pump, a CTC Analytics PAL open autosampler, an Exactive mass spectrometer (Thermo Fisher Scientific, Rockford, IL, USA) and a Poroshell 120 C18 column (150 x 3 mm, 2.7µm particle size). The column was maintained at 40°C and the flow rate was 0.5 mL/min. A gradient containing 0.1% formic acid with 20 mM ammonium acetate (A), and acetonitrile with 0.1% formic acid (B) was applied. The gradient was hold on 10% (B) for first 1 min and then increased from 10 to 100% (B) over the next 15 min, where is remained for the next 3 min. the (B) was then retained to 10% over the next 1 min, and this was followed by a 5 min re-equilibration period at 10% (B) prior to the next injection. The injection volume was 5 µL. The mass spectrometer was operated at ESI positive and negative mode. Scan range was m/z 100 – 1000. Resolution was set at 70000 (defined at m/z = 200 and was set at full width at half maximum). C-trap was set to allow 10^6 charges and the maximum injection time was 100 ms.

General Sample Preparation and Data Analysis

Candies, preserved fruits and beverages were sampled from local markets. Homogenized sample (1 ~ 5 g) each was added of 1:1 methanol/water (v/v) 25 mL in a 50-mL centrifuge tube with a ceramic grinder. The centrifuge tube was then sharked at 1000 rpm using a Geno/Grinder (SPEX SamplePrep, Metuchen, NJ, US). The extract was diluted to 50 mL and then centrifuged and filtered with a 0.22 µm PVDF filter into an LC vail for analysis. Data were evaluated by the software ToxID 2.1.1 (Thermo Fisher Co.). The target analytes were extracted by permitting a mass window of 10 ppm and matched RT.

Detection of Amaranth in Syrup-Preserved Olive

The qualitative analysis was based on calculated elemental composition of potential analytes. For example, Amaranth has elemental formula $C_{20}H_{14}N_2O_{10}S_3$, the calculated monoisotopic mass was 536.97388. The software automatically screened a total ion chromatogram (TIC) and extracted matched ions form the detection list (Table **2**). The accurate mass and retention time was used for identification. The screening results were showed in Fig. (**1**) for a syrup-preserved olive sample. E102 (Tartrazine), Amaranth, E124 (Ponceau 4R), Light Green SF, and E133 (Brilliant blue FCF) were detected in the sample.

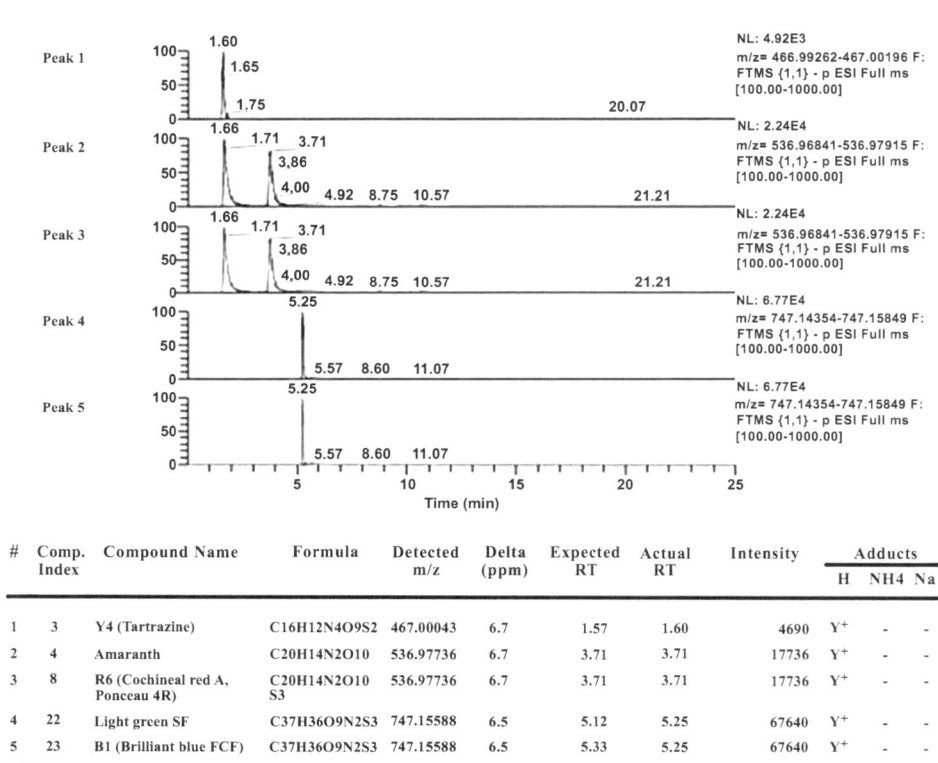

#	Comp. Index	Compound Name	Formula	Detected m/z	Delta (ppm)	Expected RT	Actual RT	Intensity	Adducts		
									H	NH4	Na
1	3	Y4 (Tartrazine)	C16H12N4O9S2	467.00043	6.7	1.57	1.60	4690	Y^+	-	-
2	4	Amaranth	C20H14N2O10	536.97736	6.7	3.71	3.71	17736	Y^+	-	-
3	8	R6 (Cochineal red A, Ponceau 4R)	C20H14N2O10 S3	536.97736	6.7	3.71	3.71	17736	Y^+	-	-
4	22	Light green SF	C37H36O9N2S3	747.15588	6.5	5.12	5.25	67640	Y^+	-	-
5	23	B1 (Brilliant blue FCF)	C37H36O9N2S3	747.15588	6.5	5.33	5.25	67640	Y^+	-	-

Fig. (1). Screening result of a syrup-preserved olive sample by ToxID softwave based on an established detection list.

Light Green SF and E133 were isomers sharing the same elemental composition, hence, were positively detected by the software. However, combining the retention time and accurate monoisotopic mass, the unknown dye had nearest RT to E133. It was identified as E133. Amaranth and E124 were isomers too, they co-eluted at RT near 3.17 min. These two compounds required further identification. Fig. (**2**) showed the mass spectra of fragments for Amaranth, E124, and an unknown compound from sample solution. Apparently, Amaranth was confirmed by matched mass spectrum. Amaranth was a sulfonic acid-based naphthylazo dye, banned by the U.S. Food and Drug Administration (FDA) in 1976 for use in foods, drugs, and cosmetics due to suspected carcinogenic activity [12].

Table 2. The semi-targeted detection list of sweetens, preservatives and dyes by LC/Orbitrap MS.

No.	Compound	Elemetal Formula	Charge	RT (min)
1	Saccharin	$C_7H_5NO_3S$	-	2.70
2	Cyclamate	$C_6H_{13}NO_3S$	-	3.26
3	Aspartame	$C_{14}H_{18}N_2O_5$	+	4.64
4	Stevioside	$C_{38}H_{60}O_{18}$	+	5.99
5	Acesulfame Potassium	$C_4H_5NO_4S$	-	2.16
6	Dulcin	$C_9H_{12}N_2O_2$	+	5.43
7	Neohesperidin dihydrochalcone	$C_{28}H_{36}O_{15}$	+	5.75
8	Glycyrrhizic acid	$C_{42}H_{61}O_{16}$	-	6.55
9	Neotame	$C_{20}H_{30}N_2O_5$	+	6.58
10	Alitame	$C_{14}H_{25}N_3O_4S$	+	4.94
11	Benzoic acid	$C_7H_6O_2$	-	5.51
12	p-Hydroxybenzoic acid	$C_7H_6O_3$	-	3.55
13	Salicylic acid	$C_7H_6O_3$	-	4.23
14	Methyl p-hydroxybenzoate	$C_8H_8O_3$	-	5.93
15	Ethyl p-hydroxybenzoate	$C_9H_{10}O_3$	-	6.79
16	Propyl p-hydroxybenzoate	$C_{10}H_{12}O_3$	-	7.68
17	iso-Propyl p-hydroxybenzoate	$C_{10}H_{12}O_3$	-	7.61
18	Butyl p-hydroxybenzoate	$C_{11}H_{14}O_3$	-	8.45
19	iso-Butyl p-hydroxybenzoate	$C_{11}H_{14}O_3$	-	8.43
20	sec-Butyl p-hydroxybenzoate	$C_{11}H_{14}O_3$	-	8.26
21	Dehydroacetic acid	$C_8H_8O_4$	+	6.35
22	Natamycin	$C_{33}H_{47}NO_{13}$	-	6.13
23	E124, Red 6 (Taiwan), Red 102 (Japan) Cochineal Red A, Ponceau 4R	$C_{20}H_{14}N_2O_{10}S_3$	-	3.32
24	E127, FD&C Red 3, Red 7 (Taiwan), Red 3 (Japan), Erythrosine	$C_{20}H_6I_4O_5Na_2$	+	7.85
25	E102, FD&C Yellow 5, Yellow 4(Taiwan, Japan), Ttartrazine	$C_{16}H_{12}N_4O_9S_2$	-	1.57

(Table 2) contd.....

No.	Compound	Elemetal Formula	Charge	RT (min)
26	E110, FD&C Yellow 6, Yellow 5(Taiwan, Japan), Sunset Yellow	$C_{16}H_{12}N_2O_7S_2$	-	3.93
27	E133, FD&C Blue 1, Blue 1(Taiwan, Japan), Brilliant blue FCF	$C_{37}H_{36}N_2O_9S_3$	-	5.21
28	E132, FD&C Blue 2, Blue 2(Taiwan, Japan), Indigo Carmine	$C_{16}H_{10}N_2O_8S_2$	+2	6.32
29	E143, FD&C Green 3, Green 3 (Taiwan, Japan), Fast Green FCF	$C_{37}H_{36}N_2O_{10}S_3$	-	4.98
30	E129, FD&C Red 40, Red 40 (Taiwan, Japan), Allura Red AC	$C_{18}H_{16}N_2O_8S_2$	-	4.08
31	E104, Quinoline yellow S	$C_{18}H_{11}NO_5S$	-	5.50
32	Xylene Fast Yellow 2G, Acid Yellow 17	$C_{16}H_{12}Cl_2N_4O_7S_2$	+	4.11
33	Alpha-naphthol orange, Acid Orange 2	$C_{16}H_{12}N_2O_4S$	-	6.34
34	Crocein Orange G	$C_{16}H_{12}N_2O_4S$	-	6.32
35	Naphthol Yellow S	$C_{10}H_6N_2O_8S$	-	3.94
36	Orange G	$C_{16}H_{12}N_2O_7S_2$	+	4.04
37	Methyl Yellow	$C_{14}H_{15}N_3$	+	11.53
38	Ponceau SX	$C_{18}H_{16}N_2O_7S_2$	-	5.06
39	Ponceau 3R	$C_{19}H_{18}N_2O_7S_2$	-	5.42
40	Erythrosin B, Red 3, Acid Red 51	$C_{20}H_8I_4O_5$	+	7.95
41	Amaranth	$C_{20}H_{14}N_2O_{10}S_3$	-	3.72
42	Rhodamine B	$C_{28}H_{30}N_2O_3$	+	8.73
43	E131, Patent Blue V, Acid Blue 1	$C_{27}H_{32}N_2O_6S_2$	+2	6.03
44	Alizarin Green, Patent Green	$C_{37}H_{35}ClN_2O_6S_2$	+	6.49
45	Alizarin, Turkey Red	$C_{14}H_8O_4$	-	7.89
46	Scarlet GN	$C_{18}H_{16}N_2O_7S_2$	+	4.70
47	Lissamine Green B	$C_{27}H_{26}N_2O_7S_2$	+	5.31
48	Azorubine, Brilliant Crimson Red, Red 10	$C_{20}H_{14}N_2O_7S_2$	-	5.09
49	Solvent Green 3	$C_{28}H_{22}N_2O_2$	+	15.36
50	Rose Bengal	$C_{20}H_4C_{14}I_4O_5$	-	9.76
51	Sudan Red 7B	$C_{24}H_{21}N_5$	+	15.01
52	Sudan I	$C_{16}H_{12}N_2O$	+	11.78
53	Sudan II	$C_{18}H_{16}N_2O$	+	13.11
54	Sudan III	$C_{22}H_{16}N_4O$	+	13.85
55	Sudan IV	$C_{24}H_{20}N_4O$	+	15.51
56	Sudan Black B	$C_{29}H_{24}N_6$	+	13.46

(Table 2) contd.....

No.	Compound	Elemetal Formula	Charge	RT (min)
57	Sudan Red G	$C_{17}H_{14}N_2O_2$	+	11.58
58	Sudan Orange G	$C_{12}H_{10}N_2O_2$	+	9.03
59	Para Red	$C_{16}H_{11}N_3O_3$	+	9.93
60	Chrysoidine G	$C_{12}H_{12}N_4$	+	6.97
61	Light Green SF	$C_{37}H_{36}N_2O_9S_3$	-	5.11
62	Sulforhodamine B	$C_{27}H_{30}N_2O_7S_2$	+	6.31

Ponceau 4R: $C_{20}H_{14}N_2O_{10}S_3$

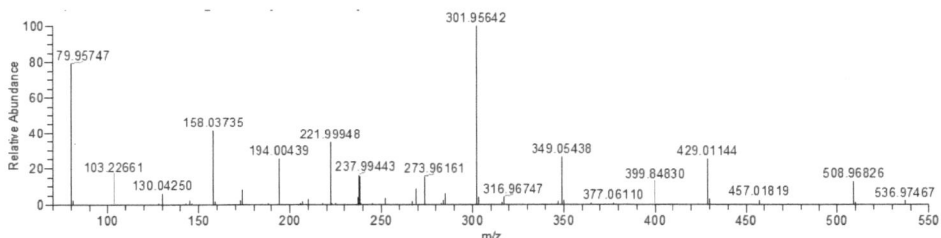

Amaranth: $C_{20}H_{14}N_2O_{10}S_3$

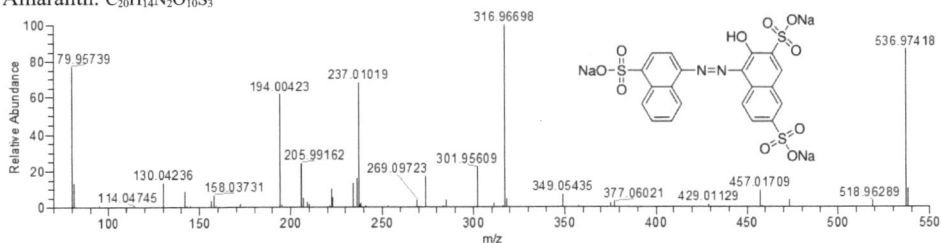

Syrup-preserved olive sample solution

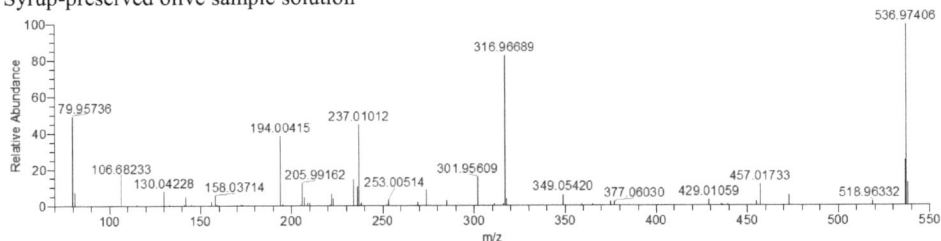

Fig. (2). Generic fragments (produced by fragmentation reaction in the MS, Higher energy collisional induced dissociation, HCD = 45V) of Ponceau 4R, Amaranth isomers, and an unknown compound from sample solution.

The TICs of a syrup-preserved olive sample were showed in Fig. (**3**). Although, several peaks were clearly appeared, none of them can indicate a single compound due to the complexity of food matrix. For example, Fig. (**4**) gave the full scan data at RT = 3.73 min. There were at least 20 compounds co-eluted simultaneously including Amaranth (m/z 536.97754). The mass spectrometer operated at full scan mode generated huge data that more than 1300 full scan spectra produced during a 20 minutes acquisition time. This technique offered retrospective data analysis approach. Data can be re-analyzed again when the detection list have been update and/or expended. The detection power can grow as long as accurate mass and RT of new compounds were added.

Since the first commercial Orbitrap Fourier transform mass spectrometry introduced in 2005 [13], high resolution and high mass accuracy enabled the elucidation of a compound formula by monitoring very narrow mass windows. If compound detected in the syrup-preserved olive remained unknown, it may be elucidated by the accurate molecular mass and the relative isotopic abundance (RIA). Fig. (**5**) gave the RIA of an analyte in syrup-preserved olive sample (top). There were three simulated elemental compositions which hit the measured mass deviation. The formula of $C_{20}H_{14}N_2O_{10}S_3$ was Amaranth (Fig. **5b**). The m/z = 536.97378 represented the monoisotope (M). The following m/z = 537.97660 represented the M+1 isotope which was most contributed by the stable isotope of 13-carbon (^{13}C) and 15-nitrogen (^{15}N) because of their relatively higher natural abundances of 1.1% and 0.4%, respectively (Table **3**). Carbon was the major component of organic compounds. The RIA of M+1 in the analyte was about 22% in Fig. (**5a**), hence, the number of carbons was estimated to be 20 based on the calculation of 22/1.1=20 (1.1 represented the natural abundance of ^{13}C). The RIA of M+2 isotope was normally assigned to the $^{13}C_2$ signal. For example, the RIA of $^{13}C_2$ in $C_{30}H_4NO_{10}$ (Fig. **5c**) was estimated to $0.011^2C_2^{30} = 5.3\%$. The RIA of M+2 in analyte in sample was estimated to be $0.011^2C_2^{20} = 2.3\%$ (Fig. **5a**), however, the measured value was about 19% which was much higher than normal. It could be considered that there were atoms other than carbon, hydrogen, oxygen, and nitrogen contributed to the M+2 isotope. Sulfur and chlorine were the best hits due to their high natural abundances of stable isotopes of 34-sulfur (^{34}S) and 37-chlorine (^{37}Cl), respectively. The natural abundance of ^{37}Cl was about 24% providing a very special M+2 pattern of molecular contained chlorine atoms (Fig. **5d**). Apparently, sulfur was the best explanation. More information on the strategy for the elucidation of elemental compositions based on VIA by high-resolution mass spectra of unknown was reported by Kaufmann [14].

Table 3. Isotopic mass of some stable isotopes with relatively higher natural abundances.

Stable isotope	Natural abundance		Isotope mass
^{13}C	1.109	%	13.003355
^2H	0.0115	%	2.014102

(Table 3) contd.....

Stable isotope	Natural abundance		Isotope mass
^{17}O	0.0373	%	16.999132
^{15}N	0.366	%	15.000109
^{33}S	0.76	%	32.971459
^{34}S	4.29	%	33.967867
^{35}Cl	75.76	%	34.968853
^{37}Cl	24.24	%	36.965903

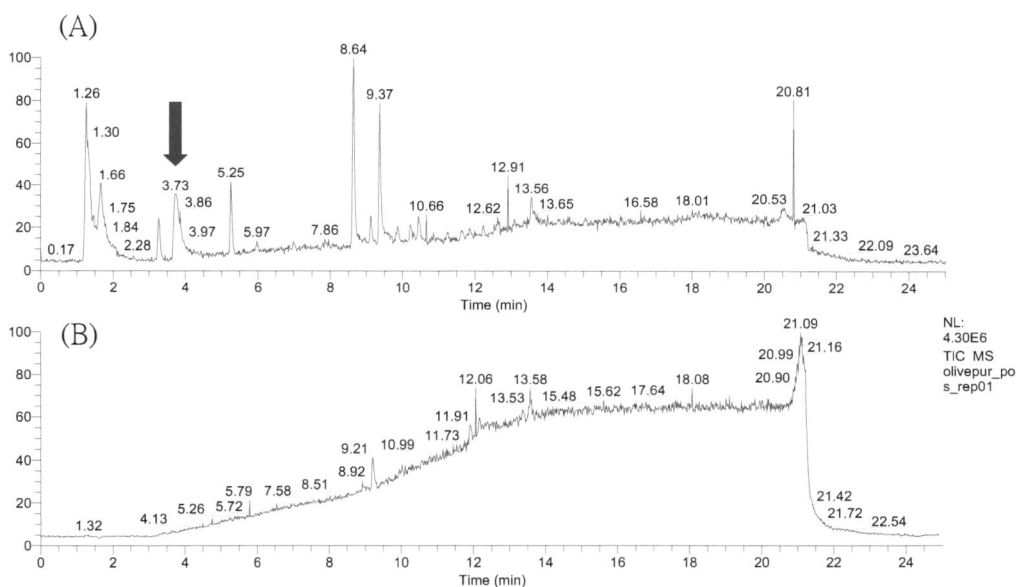

Fig. (3). Chromatograms of a syrup-preserved olive sample acquired at ESI negative charge (A), and positive charge (B).

For more broad application, Kaufmann and his coworkers reported a called semi-targeted approach which 116 compounds with no reference materials were set as detection list. Where only the accurate mass was utilized for identification. Because lack of the retention times of the investigated compounds, any signal appearing within the chromatogram could potentially be the investigated analyte and produced numerous false positive signals which required further investigations [15].

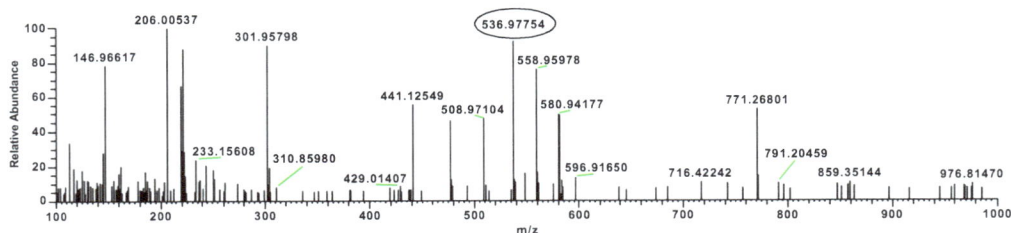

Fig. (4). Mass spectrum obtained at RT=3.73 minutes.

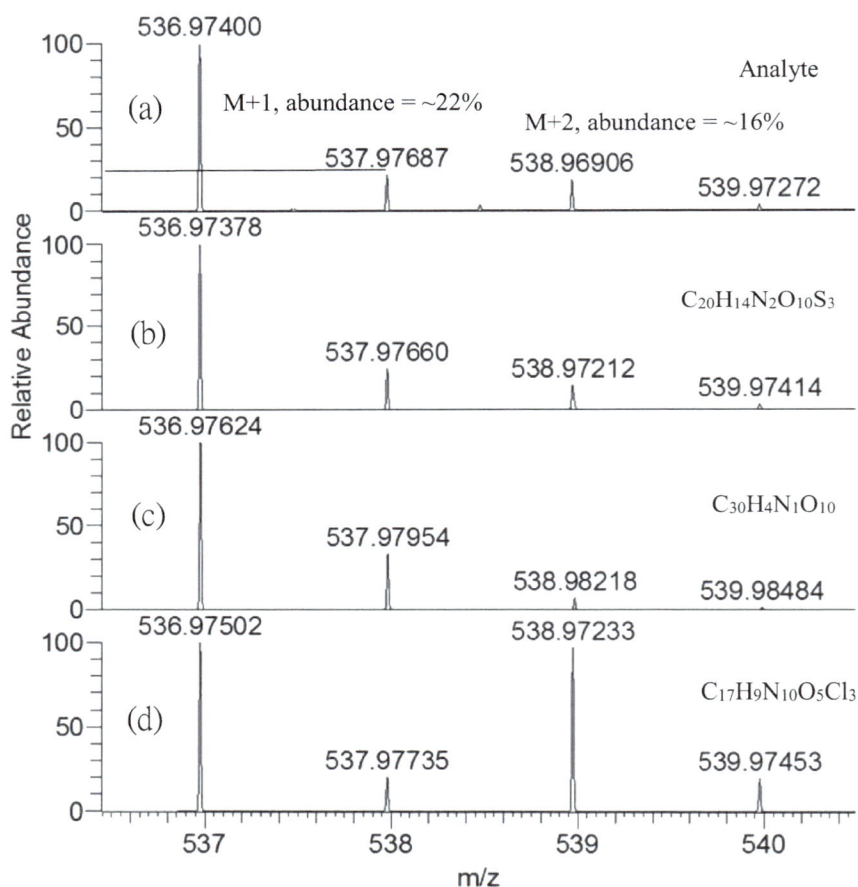

Fig. (5). Isotope pattern of a peak m/z = 536.97400 at 3.71 min in a sample (top) and three simulated signals. Permitted were C; H; N; O; S and Cl atoms. All spectra were normalized along y axis to permit of the (M + 1), (M + 2), and (M + 3) isotope abundances.

DIFFICULT SAMPLES: SAMPLE PREPARATION FOR A FISH CAKE PRODUCT AND THE DETECTION OF SULFORHODAMINE B

A reddish fish cake product was found very difficult to extract its red color by methanol/water (1:1, v/v) solution. The general sample preparation procedure can not apply. A new method modified from Harp [16] was utilized. The sample 5 g was grinded by a blender and then placed in a 50-mL centrifuge tube, followed by 25mL 1:1 methanol/water contained 10% aqueous NH_4OH and a ceramic grinder. The centrifuge tube was then sharked at 1000 rpm using a Geno/Grinder (SPEX SamplePrep, Metuchen, NJ, US). The mixture was centrifuged and filtered with No. 1 filter paper. The extract was evaporated under vacuum at 50°C to dry. The residue was dissolved in 10 mL acetonitrile (ACN) to precipitate protein, and then centrifuged and filtered. The solvent was evaporated and then re-dissolved in 5 mL ACN. A 2 mL portion of the supernatant was taken and filtered with a 0.22 μm filter into an LC vial for analysis after centrifuging.

Sulforhodamine B was positively detected by ToxID software. The chromatogram is shown in Fig. (**6**). The target was identified by matched RT and accurate mass with 2.5 ppm windows. The mass accuracy or called mass error in parts per million (ppm) was calculated through the following formula.

Mass error = exact mass − accurate mass (theoretical mass)

$$\text{Mass error in parts per million (ppm)} = \frac{\text{mass error}}{\text{exact mass}} \times 10^6$$

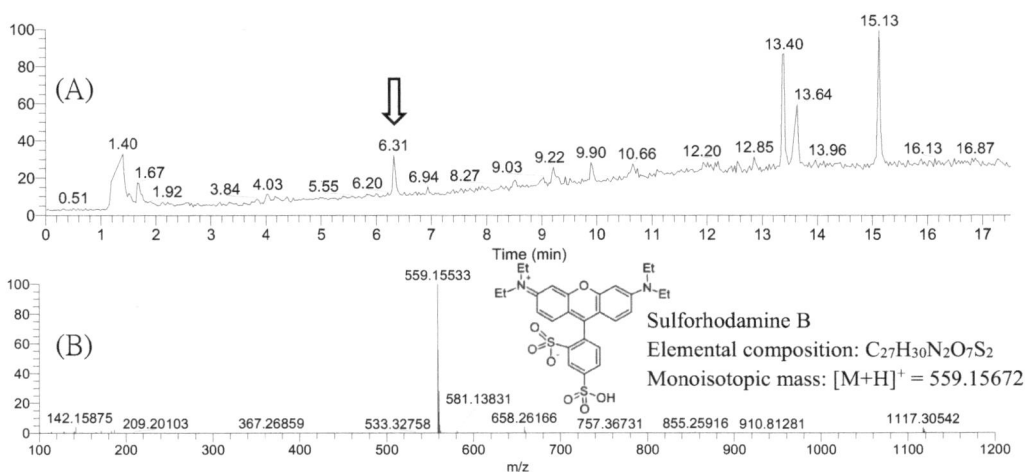

Fig. (6). Chromatogram (TIC) of a fish cake product (A), and the mass spectrum at RT = 6.31 min (B) which was suggested to be Sulforhodamine B based on the matched RT and accurate mass (within 2.5 ppm window).

SAMPLE PREPARATIONS LEAD ARTIFACTS: DETECTION OF AMINOCARMINIC ACID AND CARMINIC ACID IN A CHOCOLATE CANDY

Food colors are usually divided into natural or natural-identical (synthetically prepared) and synthetic colors [17]. Carminic acid, the major component of E120 is a red glucosidal hydroanthrapurin which belongs to natural color in EU, US, and Taiwan. It is obtained from the extract of Cochineal insects. Carmine acid changes its color from orange to red depending on the pH of the solution. Aminocarminic acid (acid-stable carmine), which is synthesized from carminic acid by heating with NH_4OH, is always purple-red against pH (Sugimoto *et al.*, 2001). Hence, Aminocarminic acid is more preferred to be used as red coloration for foods. However, Aminocarminic acid is forbidden in most of the countries including EU and Taiwan. The structures of carminic acid (E120) and Aminocarminic acid are shown in Fig. (**7**) [18].

(A) Carminic acid (E120)
   Elemental composition: $C_{22}H_{20}O_{13}$
   Monoisotopic mass: 491.08311 [M-H]⁻

(B) Aminocarminic acid
   Elemental composition: $C_{22}H_{21}NO_{12}$
   Monoisotopic mass: 490.09910 [M-H]⁻

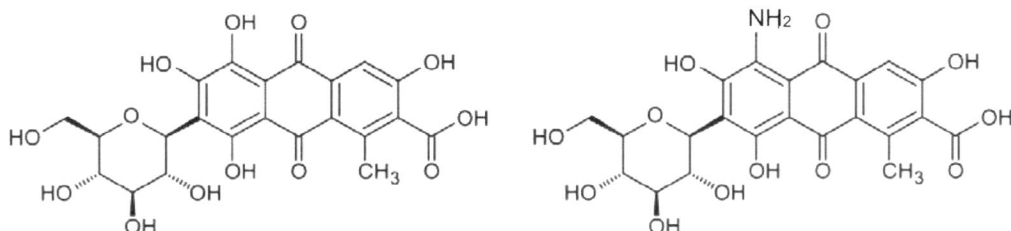

Fig. (7). Structures of Carminic acid (A) and Aminocarminic acid (B).

An imported red chocolate candy containing carminic acid as an ingredient was surveyed from border by northern center of Taiwan Food and Drug Administration (TFDA). The sample was prepared by modified method similar to fish cake as mentioned earlier. The chocolate was treated by 1:1 methanol/water containing 10% aqueous NH_4OH, followed by evaporation under vacuum at 50°C to dry. The chromatogram is shown in Fig. (**8**) with two compound identities confirmed by LC/HRMS through fragmentation (Fig. **9**).

Carminic acid was identified by the RT, molecular ion (accurate mass) and its fragments compared to reference standard. Aminocarminic acid was identified by the accurate mass and fragments referred to literature [19]. Aminocarminic acid was found in sample solution in high ratio compared to carminic acid, it seemed not conforming to the Taiwan regulation on food additives.

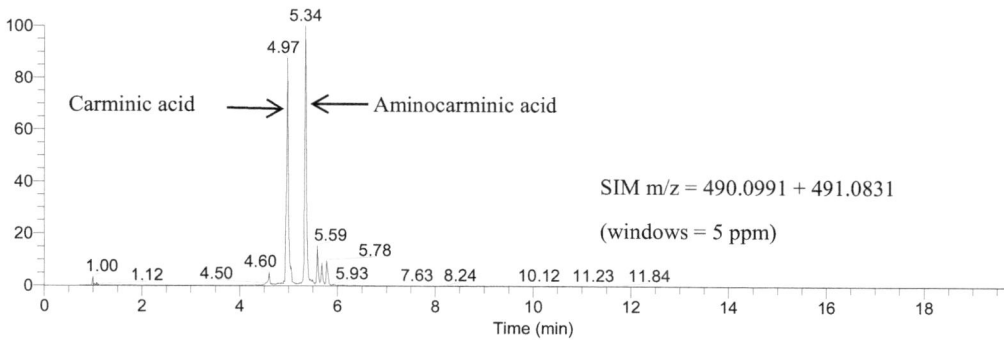

Fig. (8). Chromatogram of a chocolate candy in the selective ion monitoring (SIM) mode.

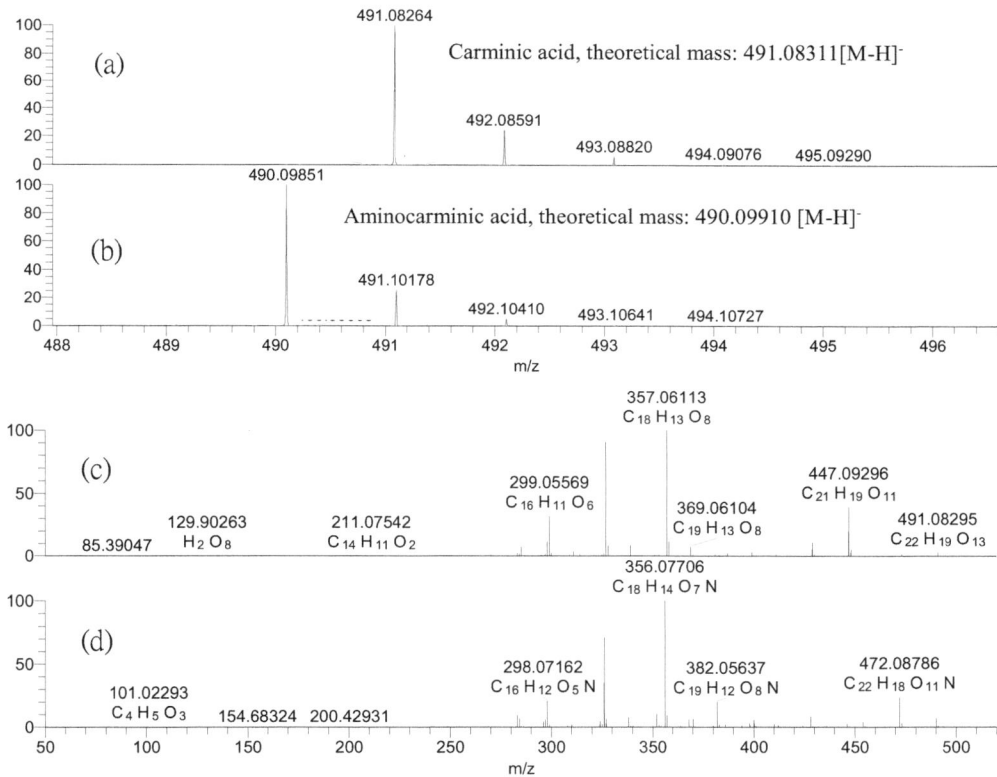

Fig. (9). Accurate mass and isotopic pattern of Carminic acid (a), and Aminocarminic acid (b). The fragments of Carminic acid (c), and Aminocarminic acid (d) at high energy collision dissociation (HCD) = 45V.

However, after further investigation of this sample, we found that the aminocarminic acid was artifact generated during sample preparation, probably form the treatment of 1:1 methanol/water contained 10% aqueous NH_4OH and the heating during evaporation. The chromatogram of chocolate sample (Fig. **10**), which prepared by 1:1 methanol/water extraction without treatment of NH_4OH and heating, indicated carminic acid was the main content of added color additive, and only trace amount of aminocarminic acid can be found. The results demonstrated unappropriated sample preparation may lead artifacts. Special care should be taken to prevent false positive, especially, NH_4OH was often applied to release colors form sample in some methods described for certified color additives [16].

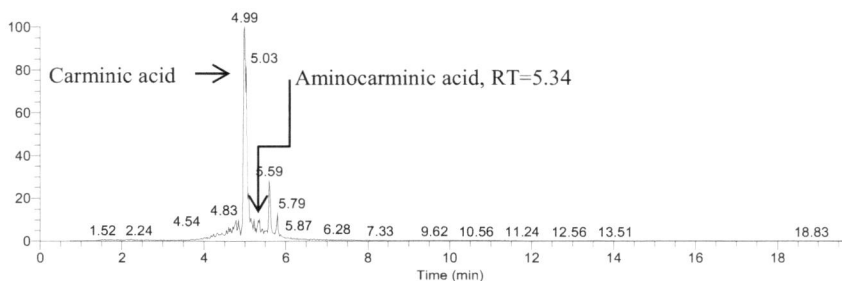

Fig. (10). The chromatogram of chocolate sample without treatment of NH_4OH and heating during sample preparation.

LEGAL OR ILLEGAL: DETECTION OF SUNSET YELLOW, CROCEIN ORANGE G AND ORANGE II IN A CANDY

In Taiwan, there are eight certified dyes (synthetic dyes) allowed to be used in food, drug, and cosmetic. The criteria of purities and impurities of certified dyes are required and listed in Taiwan regulation on additives as well in other countries. The required purities in most of certified dyes are 85%. For some natural colors the required purities may less, such as carmines (E121) which required 50% of Carminic acid in dry basis.

Under the regulation of Joint FAO/WHO Expert Committee on Food Additive (JECFA) and Taiwan as well, certified dyes are allowed certain impurities, for example, up to 5% subsidiary coloring matters are allowed in Sunset Yellow FCF. The required purities and allowed impurities of certified dyes regulated by Taiwan and JECFA are listed in Table **4**. However, the 5% impurities in Sunset Yellow FCF may be some carcinogenic, toxic, prohibited dyes or other chemicals. It is very confusing and hard to define its legality. For example, a food contained 200 ppm Sunset Yellow FCF and 10 ppm Sudan Red, if the manufacture claimed the Sudan Red was impurity of Sunset Yellow FCF.

The following is an example that two illegal dyes Crocein Orange G (Acid Orange 12) and Orange II (Acid Orange 7) along with a legal dye Sunset Yellow FCF were detected in an imported candy product. Fig. (**11**) showed the chromatogram of a candy sample, data was acquired at full scan mode, and then processed by retrospective data analysis at extracted ion mode (EIM) m/z 407.00135 which represented Sunset Yellow FCF and EIM m/z 327.04450 which represented two impurities, Crocein Orange G and Orange II (Crocein Orange G and Orange II are isomers, RT = 7.49). Fig. (**12**) (a) gave the fragments of impurities, and the fragments of standards of Crocein Orange G and Orange II were showed in Fig. (**12**) (b) and (c), respectively. Apparently, Crocein Orange G and Orange II were confirmed in the candy sample.

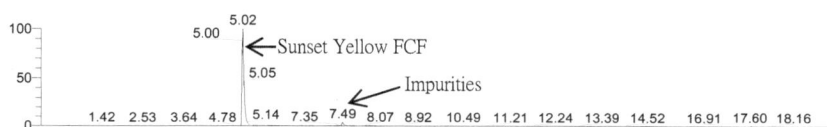

Fig. (11). Chromatograms of a candy sample. Retrospective data analysis on extracted ion mode (EIM) m/z 327.04450 and m/z 407.00135 (window = 10 ppm).

Table 4. Required purities and allowed impurities of certified color additives regulated by Taiwan and Joint FAO/WHO Expert Committee on Food Additives (JECFA).

Taiwan	FD&C	E No.	Name	Assay		[a]Subsidiary coloring matters		[b]Organic compounds other than coloring matters	
				JECFA	Taiwan	JECFA	Taiwan	JECFA	Taiwan
Blue 1	FD&C Blue No. 1	E133	Brilliant Blue FCF	85%	85%	6%	5%	0.3%	
Blue 2	FD&C Blue No. 2	E132	Indigo carminie (indogotine)	85%	85%	1%	3%	0.5%	
Green 3	FD&C Green No. 3	E143	Fast Green FCF	85%	85%	6%	5%	0.5%	
Yellow 4	FD&C Yellow No. 5	E102	Tartrazine	85%	85%	1%	3%	0.5%	
Yellow 5	FD&C Yellow No. 6	E110	Sunset Yellow FCF	85%	85%	5%	5%	0.5%	0.5%

(Table 4) contd.....

Taiwan	FD&C	E No.	Name	Assay		ᵃSubsidiary coloring matters		ᵇOrganic compounds other than coloring matters	
				JECFA	Taiwan	JECFA	Taiwan	JECFA	Taiwan
Red 6		E124	New Coccin (Ponceau 4R)	80%	82%	1%	3%	0.5%	
Red 7	FD&C Red No. 3	E127	Erythrosine	87%	85%	4%	3%	0.2%	
Red 40	FD&C Red No. 40	E129	Allura Red AC	85%	85%	3%	3%	1%	1%

a) Additional regulation may be applied for each compound. For example, the subsidiary coloring matters for Sunset yellow is not more than 5% and not more than 2% for colors other than trisodium 2-hydroxy-1-(4-sulfonatophenylzo)naphthalene-3,6-disulfonate.

b) Additional regulation may be applied for each compound. For example, the organic compounds other than coloring matters for Sunset yellow is sum of the: monosodium salt of 4-aminobenzenesulfonic acid, disodium salt of 3-hydroxy-2,7-naphthalenedisulfonic acid, monosodium salt of 6-hydroxy-2-naphthalenesulfonic acid, disodium salt of 7-hydroxy-1,3-naphthalenedisulfonic acid, disodium salt of 4,4'-diazoaminobi--benzenesulfonic acid, and disodium salt of 6,6'-oxybis-2-naphthalenesulfonic acid.

UNKNOWN COMPOUND DETECTION BY QUADRUPOLE-ORBITRAP MASS SPECTROMETRY: DISCOVERY OF DIETHYL YELLOW IN EMULSIFIER

A dye screening system consisting of a LC photodiode array coupled with a HRMS was used to detect the illegal addition of dyes in foods. An illegal dye diethyl methyl was found deliberately added into an emulsifier which to be used as a defoaming agent during soymilk curd production.

Sample Extraction

Soymilk curd and emulsifier samples 2.5 g each was ground with 20 ml acetonitrile and then ultrasonicated at 35°C for 30 min. The extract was centrifuged, filtered, and then diluted with acetonitrile before analysis.

Separation and Detection by HPLC/DAD/MS/HRMS

A LC system consisted of an UltiMate 3000 Standard LC (Thermo Fisher Scientific Inc., Waltham, MA, USA), an Acquity UPLC HSS T3 column (2.1 × 100 mm), a photodiode array detector (DAD-3000RS, Thermo Fisher Scientific), and a Q Exactive Orbitrap high-resolution mass spectrometer (Thermo Fisher Scientific) were utilized. The eluent was consisted of 0.1% formic acid water solution (A) and acetonitrile with 0.1% formic acid (B). The linear gradient program was as follows: 0–1 min, 10% B; 1–11 min, 10–100% B; 11–19 min,

100% B; 19–20 min, 100–10% B; 20–25 min, 10% B. The flow rate was 250 μl min^{-1}. The photodiode array detector monitored at 415 nm with absorption spectra recorded from 350 to 750 nm was connected between a column and a HRMS.

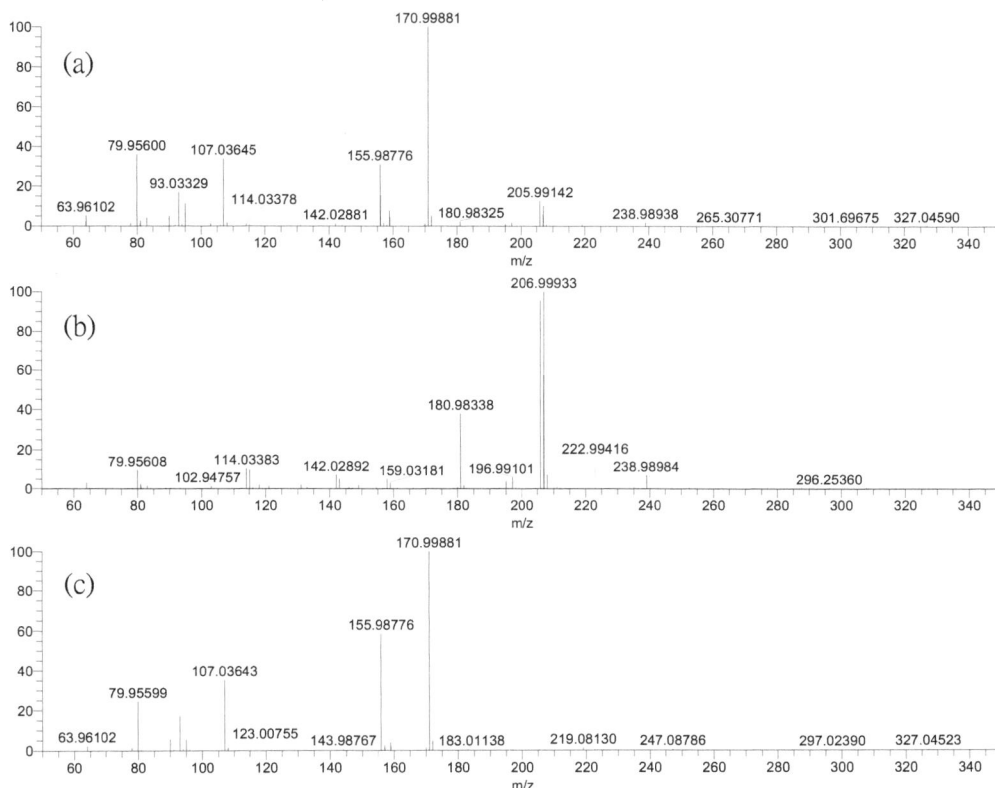

Fig. (12). High resolution mass spectra of fragments at HCD = 60V of a candy sample at RT=7.49 (a), and standard compounds of Crocein Orange G (b) and Orange II (c). Apparently, the combination of spectrum (b) and (c) represents spectrum (a).

Detection of Dimethyl Yellow and Diethyl Yellow

The coupled HPLC photodiode array (PDA) and HRMS was very useful for unknown dye screening. Mass spectrometer was a universal detector offered the analyte many advantages over specific detectors. The huge data contained accurate mass and fragments from HRMS may provide enough information for the determination of interest target analytes if there was a library for search. However, extremely complicate data offered disadvantages for unknown compound screening. For example, Fig. (**13b**) showed MS total ion chromatograms (TIC) of a yellow emulsifier sample. The chromatogram gave hundreds of peaks that make it impossible to search for unknowns by identifying peaks in one to one. The coupled PDA and HRMS offers advantages in unknown dye detection. The PDA monitored at visible 415 nm selectively pointed out only three suspicious peaks as showed in Fig. (**13a**). The first peak with RT = 5.20 min was a known compound dimethyl yellow. Followed a peak at 5.62 min was suggested an unknown yellow dye. We can located this peak (RT = 5.62 min) in Fig. (**13a**) into a MS chromatogram to find a peak at RT = 5.61 in Fig. (**13b**). The accurate monoisotopic mass of the unknown was obtained to be 254.16515. The best guess of the formula was $C_{16}H_{20}N_3$ dissolved by accurate molecular mass and RIA as described in **3.3**. The unknown was further identified as diethyl yellow (solvent yellow 56) by fragmentation and NMR spectra. This is an example of traditional PDA detector assisted modern MS detector.

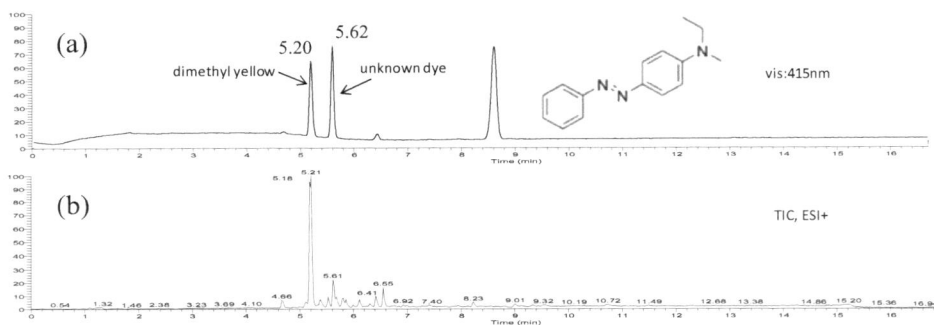

Fig. (13). Chromatograms of an unknown dye acquired by (a) visible 415 nm and (b) HRMS ESI⁻.

DEVELOPMEHNT OF DATA-INDEPENDENT ACQUISITION (DIA) FOR THE SIMULTANEOUS SCREENING AND CONFIRMATION OF DYES IN FOODS

We already talked a screening method of semi-targeted/non-targeted detection of dyes by HPLC/HRMS. The instrument was operated at full-scan mode to obtain accurate mass of molecular ions of all analytes in a sample in terms of theory. However, the main disadvantage of this method is lacking of fragment ions for confirmation.

In order to increase screening power in dyes, we established a target list of 33 dyes with fragments information showed in Table **5** by HRMS. The target list allows for data dependent acquisition just like traditional low resolution MS/MS with a selected reaction monitoring (SRM) list which is still a gold standard of analytes confirmation nowadays. However, a SRM or target inclusion list only allows signals which fit target list to be recorded. It will lose all information of semi-target, non-target or even unknown. This way limits the screening power of a method. In order to increase the possibility of semi-target and non-target detection, all molecular ions (precursor ions) and theirs fragments (product ions) in a sample have to be recorded. However, for semi-target or non-target, the bottleneck is how to select unexpected (unknown) precursor ions for fragmentation in advance in order to obtain their product ions.

Table 5. Screening target list of 33 dyes.

No	Compound	Formula	Mass	P	C	RT	Fragment	Fragment	Fragment
1	Amaranth	$C_{20}H_{14}N_2O_{10}S_3$	267.9832	-	2	6.44	228.0046	221.0149	233.9862
2	Auramine	$C_{17}H_{21}N_3$	268.1808	+	1	7.17	147.0918	252.1497	122.0965
3	Azorubine	$C_{20}H_{14}N_2O_7S_2$	228.0048	-	2	6.33	221.0148	170.0246	79.9572
4	Basic orange 2 (chrysoidine G)	$C_{12}H_{12}N_4$	213.1134	+	1	6.00	121.0636	95.0493	105.0449
5	Benzyl violte 4B (acid violet 49)	$C_{39}H_{41}N_3O_6S_2$	710.2364	-	1	7.74	630.2808	540.2313	170.0042
6	Metanil yellow (acid yellow 36)	$C_{18}H_{15}N_3O_3S$	354.0906	+	1	9.01	185.0019	169.0889	109.0286
7	Methyl yellow (butter yellow)	$C_{14}H_{15}N_3$	226.1338	+	1	11.58	95.0493	105.0449	121.0888
8	Orange II (Acid orange 7)	$C_{16}H_{12}N_2SO_4$	327.0445	-	1	7.95	170.9994	155.9886	107.0376
9	Quinoline yellow S	$C_{18}H_{11}NO_5S$	352.0285	-	1	6.78	288.0660	221.0149	170.0246
10	Rhodamine B	$C_{28}H_{30}N_2O_3$	443.2329	+	1	8.73	399.1716	413.1867	
11	Sudan I	$C_{16}H_{12}N_2O$	249.1022	+	1	12.26	232.0998	156.0446	93.0575
12	Sudan II	$C_{18}H_{16}N_2O$	277.1335	+	1	13.58	260.1312	156.0446	121.0888
13	Sudan III	$C_{22}H_{16}N_4O$	353.1396	+	1	14.32	197.0951	196.0874	156.0447
14	Sudan IV	$C_{24}H_{20}N_4O$	381.1709	+	1	15.89	224.1183	276.1133	91.0543
15	Auramine O	$C_{17}H_{21}N_3$	268.1808	+	1	7.17	147.0917	252.1495	122.0965
16	Diethyl yellow	$C_{16}H_{19}N_3$	254.1651	+	1	12.34	225.1261	120.0808	95.0492
17	Erythrosin B	$C_{20}H_8I_4O_5$	836.6623	+	1	10.92	189.0700	329.0448	200.0621
18	Para red	$C_{16}H_{11}N_3O_3$	294.0873	+	1	11.69	128.0496	156.0445	277.0849
19	Solvent green 3	$C_{28}H_{22}N_2O_2$	419.1754	+	1	15.70	401.1654	327.1129	313.1102
20	Sudan black B	$C_{29}H_{24}N_6$	457.2135	+	1	13.63	194.0839	211.1104	247.1104
21	Sudan orange G	$C_{12}H_{10}N_2O_2$	215.0815	+	1	9.71	93.0574	122.0237	198.0788
22	Sudan red 7B	$C_{24}H_{21}N_5$	380.1869	+	1	15.46	183.0918	169.0761	115.0543
23	Sudan red B	$C_{24}H_{20}N_4O$	381.1709	+	1	15.91	276.1132	224.1183	91.0543
24	Sudan red G	$C_{17}H_{14}N_2O_2$	279.1128	+	1	12.12	123.0679	108.0444	156.0444

(Table 5) contd.....

No	Compound	Formula	Mass	P	C	RT	Fragment	Fragment	Fragment
25	Sulforhodamine B	$C_{27}H_{30}N_2O_7S_2$	559.1567	+	1	7.64	515.0946	471.0320	529.1098
26	Allura red AC	$C_{18}H_{16}N_2O_8S_2$	225.0101	-	2	5.88	214.0178	206.9993	200.0021
27	Brilliant Blue FCF	$C_{37}H_{36}O_9N_2S_3$	747.1510	-	1	7.26	667.1947	561.1157	481.1597
	Brilliant Blue FCF (-2)		373.0718	-	2	7.26	333.0932	170.0040	79.9573
28	Cochineal red A	$C_{20}H_{14}N_2O_{10}S_3$	536.9735	-	1	7.15	513.2109	301.9560	248.8599
	Cochineal red A (-2)		267.9832	-	2	7.15	273.9612	270.4144	
29	Erythrosine	$C_{20}H_8I_4O_5$	834.6477	-	1	6.32	536.8490		
30	Fast green FCF	$C_{37}H_{36}O_{10}N_2S_3$	381.0693	-	2	6.94	161.0452	248.9601	296.9811
31	Indigo carmine	$C_{16}H_{10}N_2O_8S_2$	420.9805	-	1	4.44	341.0231		
	Indigo carmine (-2)		209.9804	-	2	4.44	178.0054	170.0079	79.9572
32	Sunset yellow	$C_{16}H_{12}N_2O_7S_2$	407.0013	-	1	5.32	339.1996	327.0444	206.9993
	Sunset yellow (-2)		202.9970	-	2	5.32	170.9990		
33	Tartrazine	$C_{16}H_{12}N_4O_9S_2$	466.9972	-	1	0.88	197.9863	172.0069	79.9572

Data independent acquisition (DIA) fragments all of the molecular species within a given window without preselection of a precursor ion or regardless the detection of a precursor ion within the window was first developed for an ion trap mass spectrometer [20]. We developed a HPLC Q-Orbitrap MS with DIA concept to record all possible analytes in a sample. The operation of mass spectrometer was described as Fig. (**14**). A full scan from m/z 100 ~ 900 was first initiated at resolution of 70000. Then a window of 25 was selected one after another from m/z 100 to 500 for further fragmentations (higher energy collision dissociation, HCD = 45V). Total 16 mass spectra of fragments were obtained and the resolution was set at 17500. For m/z 500 to 900, a wider window of 100 was used one after another, and 4 mass spectra of fragments were generated at resolution set at 35000. An acquisition cycle including a full scan and 20 fragmentations required 2.6 sec. A sample had to inject into HPLC twice, one for positive ESI-MS spectrum and one for negative ESI-MS spectrum.

Materials and Sample Preparation

Candies, meat snacks, taro powder and soft drinks in total of 21 samples were purchased from local markets in Taipei City. The color additives were extracted from samples by developed method (Harp *et al.*, 2013). A 5g sample was homogenized and weighted in a centrifuge tube. Methanol and 10% aqueous NH_4OH (v/v) at ratio of 7:3 was added, and the tube was shaken at 1000 rpm using a Grinder, followed sonication for 30 min. The aqueous extract was obtained by centrifuge and then cooled in a freezer for 1 hour. The tube was centrifuged and then 2 mL portion of the aqueous lower layer was mixed with 20 μL acetic acid. Extract was filtered with 0.45μm filter before analysis.

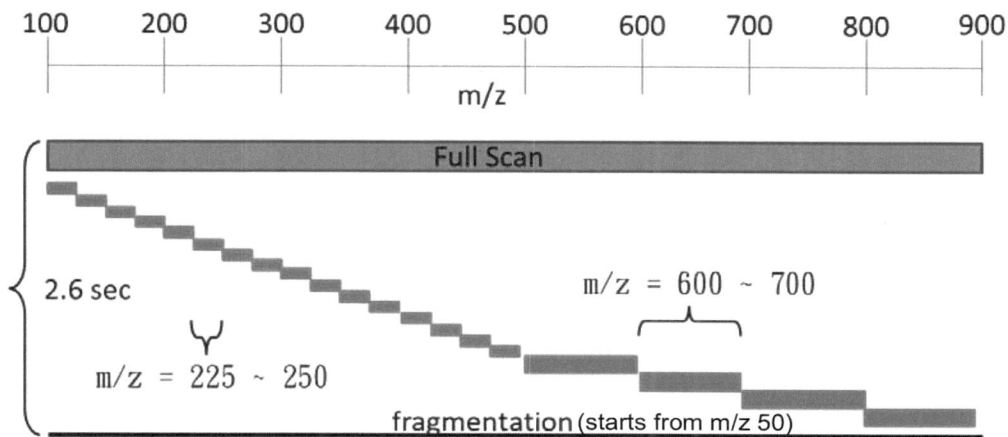

Fig. (14). Concept of data independent acquisition (DIA) in this study.

Separation

Sample extract 5 μL was injected into a LC system consisted of an UltiMate 3000 pump, an Acquity UPLC HSS T3 column (2.1 × 100 mm) and a Q Exactive Orbitrap high-resolution mass spectrometer (Thermo Fisher Scientific). The eluent was consisted of water (A) and acetonitrile (B). The linear gradient program was as follows: 0–1 min, 10% B; 1–16 min, 10–100% B; 16–24 min, 100% B; 24–25 min, 100–10% B; 25–30 min, 10% B. The flow rate was 0.25 mL/min.

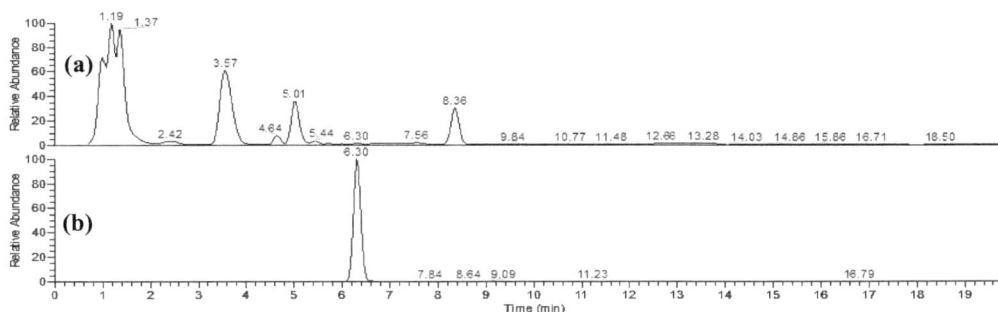

Fig. (15). Chromatograms of a soft drink sample in (a) TIC and (b) EIC (m/z 228.0048 ± 5ppm).

Evaluating the Feasibility of Applying DIA in Dye Screening

Samples were analyzed by DIA concept, and the data was processed by

TraceFinder software (Thermo Scientific Co.). In DIA, the product ions of an analyte were produced by selecting precursor ions within a wide window (m/z 25 or 100). A wide window might include other precursor ions. Therefore, DIA data was noisy and typically underwent 5 to 10 fold reduction in precursor selectivity compared to selected reaction monitoring (SRM) or multiple reaction monitoring (MRM). Table **6** shows the result of dye screening in 21 samples by DIA. There are 9 dyes found in 6 out of 21 samples. An illegal dye Azorubine was detected in a soft drink sample. Fig. (**15**) shows the total ion chromatogram (TIC) and extracted ion chromatogram (EIC) of the soft drink sample acquired by DIA through negative electrospray ionization (ESI-). The m/z 228.0048 indicated the theoretical monoisotopic of Azorubine and the retention time (RT) 6.30 matched the screening data list (Table **5**). The full scan mass spectrum (m/z 100 -900) at RT = 6.30 min is shown in Fig. (**16**).

Apparently, there were at least 10 co-elutes. We can take this example and zoom in scale to look how DIA works. Fig. (**17**) showed the mass spectrum of soft drink sample at RT = 6.30 and at m/z 200 – 275. Because a window of 25 was selected one after another from m/z = 100 to 500 in DIA, there were three windows selected between m/z 200 – 275. The target Azorubine (m/z 228.0046) was selected along with other precursors (such as m/z 228.5060, 236.0020, 236.9859 and 245.0243) from m/z 225 – 250 for further fragmentation. Fig. (**18a**) gave the result of a noisy mass spectrum compared to a clean one which only m/z 228.0 ± 0.2 was selected as precursor as (Fig. **18b**). However, due to high resolution mass spectrometry, we can still pick m/z 79.9572, 170.0246, and 221.0145 for the confirmation of Azorubine. The other mass in Fig. (**18a**) were fragments from other molecular species within m/z 225 - 250 window.

Table 6. Sample information and results of detection.

Product name	Color additives
Chocolate	Brilliant Blue FCF
Pork jerky	Sunset Yellow FCF
Taro powder	Brilliant Blue FCF, Erythrosine, Allura Red AC
Taro flavor	Brilliant Blue FCF, Allura Red AC
Fish snack	Sunset Yellow FCF
Soft drink	Azorubine

DIA generated multiplexed data allowed retrospective detections. In this study, 21 samples were tested for 33 dyes screening (Table **5**), and 9 dyes were found in 6 samples by matched molecular ion (< 5 ppm), RT and two fragments (< 10 ppm) for each analyte. We modified the matching criteria, for example, only matching molecular ions and RTs of analytes in samples with screening list (Table **5**). The results are shown in Fig. (**19**). Forty five presumptive positives were obtained, apparently, most were false positive.

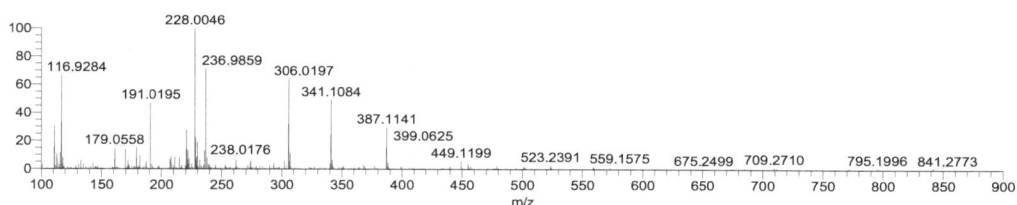

Fig. (16). A full scan mass spectrum (m/z 100 – 900) at RT = 6.30 min of a soft drink sample.

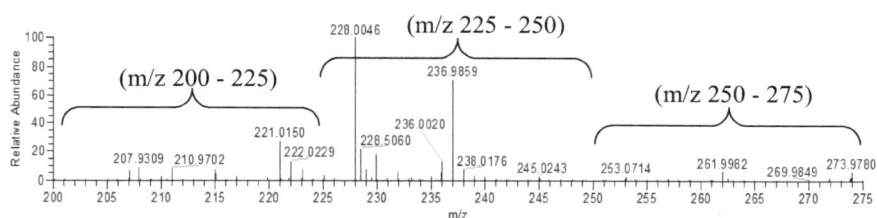

Fig. (17). Mass spectrum of soft drink sample at RT = 6.30 and at m/z 200 – 275. Target analyte (Azorubine, m/z 228.0046) was sent into HCD (high energy collision dissociation) along with other precursors within gave window (m/z 225 – 250).

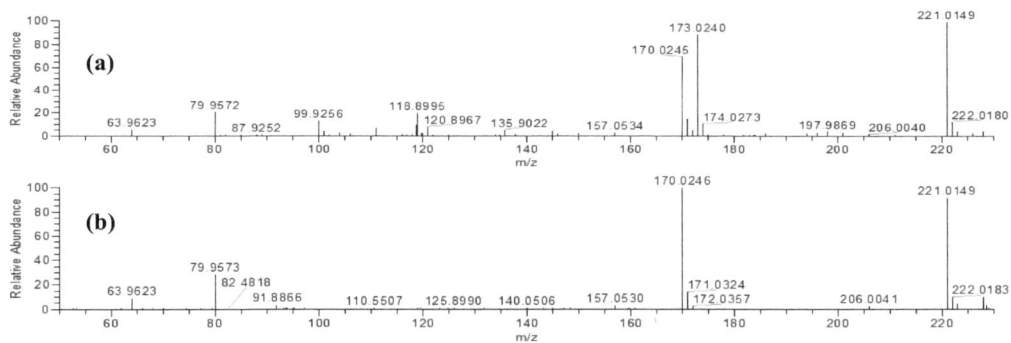

Fig. (18). Mass spectra of (a) product ions from all molecular species within a wide window (m/z 225 – 250), and (b) from precursor selected within m/z 228.0±1.0

Recently, a document released by US FDA dressed the recommended criteria on non-targeted analysis [21]. High resolving power (> 50,000) was needed when using narrower mass extraction windows (<5 ppm) in complex matrices, and presumptive positives may need further evaluation using additional product or isotope ions as well as retention time. The results in this study showed that matched RT and molecular ion (< 5 ppm) of an analyte identified by HRMS along were not sufficient. If matched RT, molecular ion (< 5 ppm) and one fragment (< 10 ppm) were set to be the confirmative criteria, 15 presumptive positives were obtained (Fig. **19**). Although these criteria met requirements for confirmation of identity [21], we found there were still not sufficient. These 15 presumptive positives were further evaluated using targeted MS/MS. The spectra of product ions were compared to contemporary standards. Only nine were matched. Therefore, we strongly suggested two fragments combined molecular ion and RT for the purpose of confirmation as showed in Fig. (**19**). Only nine dyes were screening out from 6 of 21 samples.

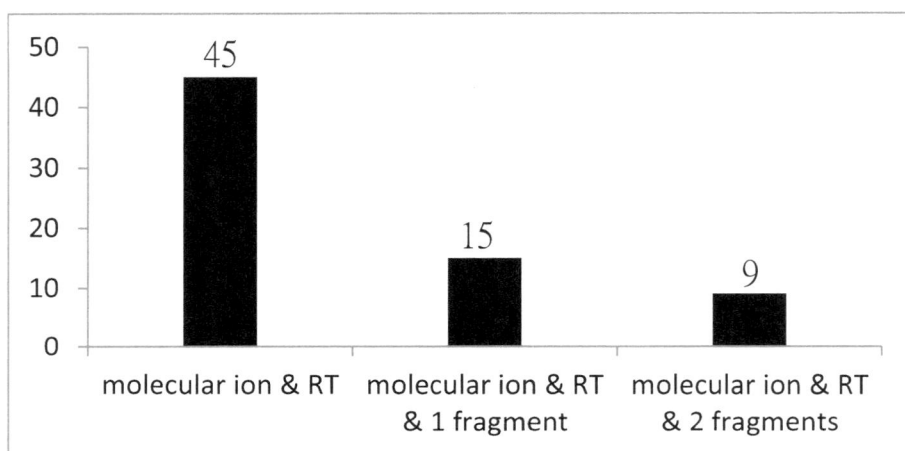

Fig. (19). Positive detections of dyes by computer software in different criteria.

CONSENT FOR PUBLICATION

Not applicable.

CONFLICT OF INTEREST

The author (editor) declares no conflict of interest, financial or otherwise.

ACKNOWLEDGNEMT

Declared none.

REFERENCES

[1]   Stiborová M, Martínek V, Rýdlová H, Hodek P, Frei E. Sudan I is a potential carcinogen for humans: evidence for its metabolic activation and detoxication by human recombinant cytochrome P450 1A1 and liver microsomes. Cancer Res 2002; 62(20): 5678-84.
[PMID: 12384524]

[2]   Seidenari S, Manzini BM, Danese P. Contact sensitization to textile dyes: description of 100 subjects. Contact Dermat 1991; 24(4): 253-8.

[http://dx.doi.org/10.1111/j.1600-0536.1991.tb01718.x] [PMID: 1831105]

[3]   He L, Su Y, Fang B, Shen X, Zeng Z, Liu Y. Determination of Sudan dye residues in eggs by liquid chromatography and gas chromatography-mass spectrometry. Anal Chim Acta 2007; 594(1): 139-46.
        [http://dx.doi.org/10.1016/j.aca.2007.05.021] [PMID: 17560395]

[4]   Qi P, Zeng T, Wen Z, Liang X, Zhang X. Interference-free simultaneous determination of Sudan dyes in chili foods using solid phase extraction with HPLC-DAD. Food Chem 2011; 125: 1462-7.
        [http://dx.doi.org/10.1016/j.foodchem.2010.10.059]

[5]   Di Anibal CV, Odena M, Ruisánchez I, Callao MP. Determining the adulteration of spices with Sudan I-II-II-IV dyes by UV-visible spectroscopy and multivariate classification techniques. Talanta 2009; 79(3): 887-92.
        [http://dx.doi.org/10.1016/j.talanta.2009.05.023] [PMID: 19576460]

[6]   Li J, Ding XM, Liu DD, *et al.* Simultaneous determination of eight illegal dyes in chili products by liquid chromatography-tandem mass spectrometry. J Chromatogr B Analyt Technol Biomed Life Sci 2013; 942-943: 46-52.
        [http://dx.doi.org/10.1016/j.jchromb.2013.10.010] [PMID: 24212142]

[7]   Tateo F, Bononi M, Gallone F. Rapid detection of dimethyl yellow in curry by liquid chromatography-electrospray-tandem mass spectrometry. Czech J Food 2010; 28: 427-32.
        [http://dx.doi.org/10.17221/135/2009-CJFS]

[8]   Xu T, Wei KY, Wang J, *et al.* Development of an enzyme-linked immunosorbent assay specific to Sudan red I. Anal Biochem 2010; 405(1): 41-9.
        [http://dx.doi.org/10.1016/j.ab.2010.05.031] [PMID: 20522332]

[9]   Di Anibal CV, Ruisanchez I, Callao MP. High-resolution ^{1}H nuclear magnetic resonance spectrometry combined with chemometric treatment to identify adulteration of culinary spices with Sudan dyes. Food Chem 2011; 124: 1139-45.
        [http://dx.doi.org/10.1016/j.foodchem.2010.07.025]

[10]   IARC Monographys on the Evaluation of the Carcinogenic Risk of Chemicals to Man. IARC 1975; vol. 8: 125.

[11]   Jia W, Chu X, Ling Y, Huang J, Lin Y, Chang J. Simultaneous determination of dyes in wines by HPLC coupled to quadrupole orbitrap mass spectrometry. J Sep Sci 2014; 37(7): 782-91.
        [http://dx.doi.org/10.1002/jssc.201301374] [PMID: 24478185]

[12]   Penman KG, Halstead CW, Matthias A, *et al.* Bilberry adulteration using the food dye amaranth. J Agric Food Chem 2006; 54(19): 7378-82.
        [http://dx.doi.org/10.1021/jf061387d] [PMID: 16968108]

[13]   Makarov A, Denisov E, Kholomeev A, *et al.* Performance evaluation of a hybrid linear ion trap/orbitrap mass spectrometer. Anal Chem 2006; 78(7): 2113-20.
        [http://dx.doi.org/10.1021/ac0518811] [PMID: 16579588]

[14]   Kaufmann A. Strategy for the elucidation of elemental compositions of trace analytes based on a mass resolution of 100,000 full width at half maximum. Rapid Commun Mass Spectrom 2010; 24(14): 2035-45.
        [http://dx.doi.org/10.1002/rcm.4612] [PMID: 20552696]

[15]   Kaufmann A, Butcher P, Maden K, Walker S, Widmer M. Semi-targeted residue screening in complex matrices with liquid chromatography coupled to high resolution mass spectrometry: current possibilities and limitations. Analyst (Lond) 2011; 136(9): 1898-909.
        [http://dx.doi.org/10.1039/c0an00902d] [PMID: 21384037]

[16]   Harp BP, Miranda-Bermudez E, Barrows JN. Determination of seven certified color additives in food products using liquid chromatography. J Agric Food Chem 2013; 61(15): 3726-36.
        [http://dx.doi.org/10.1021/jf400029y] [PMID: 23528012]

[17]   Gennaro MC, Abrigo C, Cipolla G. High-performance liquid chromatography of food colours and its

relevance in forensic chemistry. J Chromatogr A 1994; 674(1-2): 281-99.
[http://dx.doi.org/10.1016/0021-9673(94)85234-0] [PMID: 8075774]

[18]   Sugimoto N, Kawasaki Y, Sato K, *et al.* Structure of acid-stable carmine. Shokuhin Eiseigaku Zasshi 2002; 43(1): 18-23.
[http://dx.doi.org/10.3358/shokueishi.43.18] [PMID: 11998314]

[19]   Sabatino L, Scordino M, Gargano M, *et al.* Aminocarminic acid in E120-labelled food additives and beverages. Food Addit Contam Part B Surveill 2012; 5(4): 295-300.
[http://dx.doi.org/10.1080/19393210.2012.719551] [PMID: 24786412]

[20]   Bern M, Finney G, Hoopmann MR, Merrihew G, Toth MJ, MacCoss MJ. Deconvolution of mixture spectra from ion-trap data-independent-acquisition tandem mass spectrometry. Anal Chem 2010; 82(3): 833-41.
[http://dx.doi.org/10.1021/ac901801b] [PMID: 20039681]

[21]   United States Food and Drug Administration (US FDA). Acceptance criteria for confirmation of identify of chemical residues using exact mass data within the office of foods and veterinary medicine. 2015.

CHAPTER 4

Adulteration Analysis of Honey (2014-2017)

Alankar Shrivastava*

Department of Pharmaceutical Quality Assurance, Lloyd School of Pharmacy, Greater Noida, Knowledge Park II, Greater Noida, Uttar Pradesh 201310, India

Abstract: Honey is one of the popular foods used to enhance the taste in many food articles worldwide. The importance of honey is not restricted to the boundaries of any certain countries. Honey is also important from the religious point of view. Honey is also a good source of energy. This chapter deals with the adulteration analysis of honey within last three years. This chapter describes briefly nutritious benefits and composition of honey. The analysis of adulteration of honey is described in different sections; Liquid Chromatography (LC), Nuclear Magnetic Resonance (NMR), Near Infrared Spectroscopy (NIS), Rheology and Sensors.

Keywords: Adulteration, Chromatography, Contamination, Detectors, Food, Honey, NMR, NIR Spectroscopy, Rheology, Sensors.

INTRODUCTION

Food adulteration is not a new problem and arises where there is a challenge between the physical availability and the market demand for a food item [1].

In the year 1820, Frederick Accum (German Chemist) published a book "*A Treatise on Adulterations of Food and Culinary Poisons*" [2]. His work revolutionized the field of adulteration analysis. Further development in this field leads to the current laws related to the adulteration prohibition in the food worldwide.

In order to get higher prices of worthless products manufactures are attempting to alter their products since ancient times. Consumption of safe and quality food needs quality control at each level production, storing, marketing and consumption [3].

Contamination and adulteration are two closely related terms but the former is

* **Corresponding author Alankar Shrivastava:** Department of Pharmaceutical Quality Assurance, Lloyd School of Pharmacy, Greater Noida, Knowledge Park II, Greater Noida, Uttar Pradesh 201310, India; Tel: +91-7351002560; E-mail: alankarshrivastava@gmail.com

technically unavoidable and an unintentional activity. On the other hand, adulteration is for economical or ideological gain including intentional replacement of any ingredient of the food [4].

Publication of Deelstra H. and his coworkers [5] highlighted history of food adulteration. Food adulteration and fraud attempts at their control have a long history. Whenever the supply of food increases slowly compared to urban population, the adulteration of foods becomes a problem for analytical chemists and control authorities, when food supplies increase globally at a slower rate than urban populations in many countries. The modern food laws in current practice of providing protection to the consumers are development of legal and analytical control measures adopted during middle ages.

The three main causes of food adulteration are:

1. Urge of traders for making money dishonestly.
2. Loopholes in Food Adulteration Act.
3. Wrong buying practices of consumer by ignoring their right and responsibilities towards food adulteration [6].

There are many literatures defining "honey". The recent literature by Roshan AA *et al*. [7] defines honey as:

"*the natural sweet substance produced by Apis mellifera bees from the nectar of plants, or from secretions of living parts of plants, or excretions of plant-sucking insects on the living parts of plants, which the bees collect, transform by combining with specific substances of their own, deposit, dehydrate, store and leave in the honeycomb to ripen and mature*".

Some description about its composition is given by Verzeraand Condurso (2012) [8].

"*Honey is composed mainly of monosaccharides (fructose and glucose), lesser amounts of water and a great number of minor components such as organic acids, oligosaccharides, enzymes, vitamins, minerals, pigments, a wide range of aroma compounds, and solid particles derived from honey collection.*"

Honey is described as one of the valuable food of the ancient times that delight's consumers by its various medicinal characteristics [9].

Although the food ranking system did not qualify honey as a dense source of traditional nutrients apart from the sugar content, it did emerge as a veritable source of vitamin B2, vitamin B6, iron and manganese [10].

According to the first US database on compiled information on risk factors for food fraud there are seven foods targeted for the food fraud including olive oil, milk, honey, saffron, orange juice, coffee and apple juice [11].

Honey is adulterated illegally and unethically with any other sweetening agent and sold with labeled as pure honey for financial gain or competitive advantage. This kind of advantage is known as economic adulteration [12].

The primary components of honey are sugar and water. Other major components include vitamins, aminoacids, enzymes and minerals. Although upto 200 different components reported in the available literature. The amount of dry matter varies from 95-99%. Honey shared an important part of traditional medicines from centuries ago. There are documents showing the usage of honey in gastrointestinal, cardiovascular and liver problems since ancient times. Many ancient civilizations were using honey for intestinal wound and diseases like Assyrians, Egyptians, Chinese, Greeks and Romans [13].

Honey is known to have wound healing and antibacterial property. This is due to its high viscosity causing inhibition of infection deep into the tissues and release of low level hydrogen peroxide is the reason of its antibacterial action. Its low acidity adds on its antibacterial action. Also honey, because of its flavor, color and sweetness, could be used as an ingredient or preservative in foodstuffs; for example, it has been shown that honey is able to prevent lipid peroxidation in meat [14].

Because of its several medicinal and nutritional values, a new branch of alternative therapy in the recent years have been developed offering the treatment by honey and other related products called as apitherapy [15].

A recent research by Mijanur Rahman M *et al*. 2014 [16] reported improvement in deficits in memory by polyphenols of honey. This component of honey quench biological reactive oxygen species that cause neurotoxicity, aging as well as the pathological deposition of misfolded proteins, such as amyloid beta. Polyphenol constituents of honey counter oxidative stress by excitotoxins, such as kainic acid and quinolinic acid, and neurotoxins, such as 5-*S*-cysteinyl-dopamine and 1-methyl-4-phenyl-1,2,3,6-tetrahydropyridine. Honeys are classified as unifloral or monofloral and multifloral or polyfloral [17]. Monofloral honey is generally not blended by the bees and thus limited quantity is available in the commercial market. For example, clover, alphalfa, tupelo, gall berry and cotton honey are monofloral honeys [18]. Unifloral honeys like Acacia or Orange have premium prices compared to blended or multifloral honeys [19]. In the case of monofloral honey, there should be more than 45% of the pollen collected from the derived plant species. However, this percentage may vary to more than 90%, *e.g.* chestnut honey or decreased to 10-20% for citrus, arbutus, lavender, thymus and rosemary honey. Polyfloral honey refers to the type of honey derived from different plant pollen grains [20].

There are three physical forms of honey; pressed, centrifuged and drained honey. Also, there are five different styles *i.e.* comb, chunk, crystallized or granulated, creamed and heat processed honey [21].

Tables **1** and **2** highlighted the composition of honey.

Table 1. Average composition of honey [19].

Component	Average value in %
Water	17
D-Fructose	38
D-Glucose	32
Sucrose	1.3
Maltose	7.3
Oligosaccharides	1.5
Protein	0.3
Minerals	0.2
Vitamins & amino acids	1.0

Table 2. Compositional criteria described [19].

Parameter	Value
Acidity	not more than 40%

Apparent reducing sugar, calculated as invert sugar
     Blossom honey
     Honeydew honey and blends
	
	not less than 65%
	not less than 60%

Apparent sucrose content
     In general
     Honeydew honey and blends and some special honeys
	
	not more than 5%
	not more than 10%

HMF content	not more than 40 mg/kg (ppm)

Diastase activity (Schade scale)
     In general
     Honeys with low natural enzyme content (*e.g.* citrus honeys) and an HMF content of not more than 15 mg/kg
	
	not less than 8
	not less than 3

Mineral (ash) content
     In general
     Honeydew honey and blends
	
	not more than 0.6%
	not more than 1.2%

(Table 2) contd.....

Parameter	Value
Moisture content	
In general	not more than 21%
Heather honey and clover honey	not more than 23%
Industrial honey or baker's honey	not more than 25%
Water insoluble solids content	
In general	not more than 0.1%
Pressed honey	more than 0.5%

The limited availability and high price of honey have provided a heightened interest in its adulteration [22]. Cordella *et al.*, 2005 [23], in their publication described honey adulteration as:

"the purposeful act by unscrupulous producers to incorporate sugar syrups into the natural product. In addition to nutritional and organoleptic consequences, honey adulteration can have a significant economic impact."

In the last three years (2014-16), there were many publications about evaluation of adulterated honey. In this chapter, the published literature is classified on the basis of basic principle of adulteration determination of honey.

Liquid Chromatography

Mixing of low cost honey with high cost honey to increase its value is well evident. Wang J *et al.* (2014) [24] published liquid chromatography–electrochemical detection (LC-ECD) method for the determination of low cost honey (acacia honey) with high cost honey (rape honey) at different levels (5–50%, w/w). Chromatographic conditions are described under (Table **3**).

Acacia honey is detained from blossoms of *Robinia pseudacacia*. It has a milder taste as compared to other varieties. The color varies from transparent to light yellow and does not crystallize. On the other hand, Rape honey is of light amber color and crystallizes easily. It has a flavor of rape flower and generally higher yield as compared to Acacia honey. This is the reason that Acacia honey is high priced than rape honey and due to color similarities, rape honey is often used in the adulteration of acacia honey in the market.

Adulterations have been determined by using fingerprints of chlorogenic acid and ellagic acid. Authors further stated that chlorogenic acid and ellagic acid could be considered as possible markers of acacia and rape honeys, respectively. The content of former was found to be higher in acacia honey and latter was low in rape honey.

Table 3. Details of Liquid Chromatography methods for adulteration determination in honey.

Detection	Mobile phase	Column	Chromatographic conditions	Principle Method	Adulterant	Ref.
Electrochemical	0.1% formic acid (A) and methanol (B) (v/ v) with a linear gradient as follows: 0–10 min from 5 to 15% B, 10–20 min from 15 to 15% B, 20–25 min from 15 to 17% B, 25–30 min from 17 to 30% B, 30–50 min from 30 to 40% B, 50–60 min from 40 to 55% B, and 60–70 min from 55 to 70% B, at a flow rate of 1.0 mL min^{-1}.	C$_{18}$ (250×4.6 mm, 5.0 μm)	Flow rate of mobile phase 1.0 mL min^{-1}, Temperature: 30 ° C, ECD was operated at 0.9 V in oxidative mode, and the DADs were set at 280 and 290 nm.	Cluster and Principle Component Analysis	Rape honey (Low priced) adulterated with Acacia honey	[24]
Refractive Index (RI)	80% acetone and 20% water	Zorbax carbohydrate analysis column (5 μm and 4.6 mm × 150 mm)	Flow rate, 1.4 mL/min; injection volume, 20 μL and the column temperature was set to be 25 °C.	Rheology	Saccharose and fructose syrups	[25]
Evaporative light scattering	Gradient elution: 25% A, to 50% A from 0 to 7.0 min until 7.1 min and at 12.9 min A was 0.2% triethylamine in pure water and solvent B was 0.2% triethylamine in acetonitrile	Acquity BEH amide column (2.1 × 50 mm, 1.7 μm)	Flow rate of 0.4 mL/min, Temperature: 35 °C, Detector conditions: 40 °C drift tube temperature, cooling nebulizer mode, 35 psi gas pressure, 10 pps date rate, 200 gain factor.	Simple extraction	6 different carbohydrates	[26]

(Table 3) contd.....

Detection	Mobile phase	Column	Chromatographic conditions	Principle Method	Adulterant	Ref.
Isotope Ratio Mass Spectrometer (IRMS)	Water	Hyper REZ XP Carbohydrate Ca^{2+} (300 mm × 8 mm, Phenomenex Co., USA)	Flow rate: 350 µL/min., Temperature: 65 ° C	ANOVA, Duncan's least significant test and least square method	Syrups obtained by C4 plants	[29]
Refractive index detector (RID)	Water	Carbomix Ca-NP5:8% column (7.8×300mm, 5µm)	Flow-rate was 0.3ml/min, Temperature: 80 ° C	Height of syrup indicator peak	Starch syrup	[30]
MS	Mobile phase gradient A: 30% water, B: 70% acetonitrile for the first 5 min, B reduced to 45% acetonitrile over 5 min, B adjusted to 70% over 0.1 min for 2.9 min to clean the column.	Acquity UPLC BEH Amide column (2.1 mm × 100 mm, 1.7 µm)	Flow rate of 0.3 mL/min, column temperature was maintained at 50 °C.	quadrupole time-of-light	Sugar syrups.	[31]

High Performance liquid chromatography method with Refractive Index detector for the determination of saccharose and fructose syrups as adulterants in honey is reported by Yilmaz MT and coworkers [25]. Researchers claimed that this method is not time-consuming, expensive and require moderate analytical skills. They also claimed that the method facilitates and accelerates the procedure of detection. The other advantages reported are easy cleanup procedures, less analytical skills to run the samples and economical instrumentation involved.

Zhou J *et al.* [26] in a study reported that honey adulterated with six different maltooligosaccharides (maltose, maltotriose, maltotetraose, maltopentaose, maltohexaose and maltoheptaose) was distinguished using a simple extraction procedure followed quantification by ultra performance liquid chromatography (UPLC) with evaporative light scattering detector (ESLD). This paper also highlighted the advantage of using ELS detector compared with Refractive Index (RI). The applicability of RI detector is limited because of poor S/N ratio, higher sensitivity to temperature, pressure and flow changes.

Historically, conventional gas isotope ratio mass spectrometry with manual conversion of organic samples into CO_2 gas using a sealed tube combustion method with cryogenic purification was used for the measurement of carbon stable isotopic abundance in organic samples. Usage of different inlet systems such as elemental analyzers (EA), gas chromatography (GC) and more recently liquid chromatography (LC) emerged in the last few decades [27].

EA/LC-IRMS method providing enhanced sensitivity and the ability to detect adulterations of honey with different types of sugar syrups was produced from C4 and C3 plant sources than ever before. The difference between metabolic enrichment of ^{13}C isotope values ($\Delta^{13}\delta$) of bulk honey, isolated protein, fructose, glucose, disaccharides, and trisaccharides in authentic honeys is the basic principle of this method [28].

Dong H *et al.* (2016) developed EA/LC-IRMS method in the study [29] for the identification of adulteration in commercial honey. This method first successfully provided favorable evidences in authenticity identification of honeys with C-4 sugar content of $< 0\%$, as claimed by the researchers. They found that δ^2H and $\delta^{18}O$ could potentially be helpful for the adulteration identification of commercial honey, especially, the $\delta^{18}O$ value could be a useful parameter for adulterated honey samples with C-4 sugar content (%) < -7.

Another HPLC method is proposed by Wang S *et al.* (2014) [30] for detection of honey adulteration with starch syrup and was applied in an authenticity inspection on more than 100 commercial honeys. This method in addition to its simplicity and economical procedures, the researchers recommended to be used by government organizations for the quality control of honey products.

Du B *et al.* (2015) [31] developed ultrahigh-performance liquid chromatography/ quadrupole time-of-flight mass spectrometry (UHPLC/Q-TOF-MS) method for the detection of adulterated honey. Detection markers used in this method were 2-acetylfuran-3- glucopyranoside, high-fructose corn syrup, inverted syrup, corn syrup, and rice syrup as honey adulterants; Researcher claims this method to be rapid and less time consuming as compared to other published methods and detection of many syrups (<30 min) is possible simultaneously.

Nuclear Magnetic Resonance

The intrinsic magnetic moment and angular momentum are the characteristic properties of the atoms of any molecule. Thus, on the basis of this principle NMR technique offers structural information of molecules under investigation. This technique is also known as "universal detector" because the elements such as hydrogen, oxygen, nitrogen, carbon and phosphorus, which constitute major portion in foods, have at least one isotope detectable [32]. There are several applications of Nuclear magnetic resonance (NMR) spectroscopy technique traditionally used in field related to authenticity of food. This is one of the important technique for the modern food scientists for traceability and safety of food [33]. NMR is well known powerful technique with substantial contribution in the field of food analysis. Different isotopes give rise to distinct shifted spectral lines and thus, it is possible to observe a specific isotope even in complex structures, both in solution and in solid state [34].

Application in food analysis of NMR is based on behavior (relaxation) of NMR active nuclei (*i.e.* ^1H, ^{13}C) in a magnetic field and a pulsed Radio Frequency (RF) irradiation [34]. NMR signal is generated by the emission of radio frequency (RF) because of retransmission of electromagnetic energy during relaxation. The longitudinal relaxation and transverse relation are known as T_1 and T_2 respectively [35].

The research work published by Ribeiro *et al.* (2014) described Low Field Nuclear Magnetic Resonance spectroscopy (LF 1H NMR) and physicochemical analytical methods for adulteration in honey. The level of adulteration kept was 0% (honey) to 100% (high fructose corn syrup). Researchers found increase in the relaxation time with the increase in the adulteration of honey with fructose syrup and thus concluded that this technique can be used to distinguish pure honey and high fructose corn syrup adulteration in it [36].

Another published study in the same year by Spiteri *et al.* (2014) claimed to be successfully implemented under routine conditions in the laboratory on more than 1400 commercial samples. Authenticity of both multi and mono floral honey can be checked by using H1 NMR by using suitable statistical models and quantification procedures [37].

Near Infrared Spectroscopy

The organic matter constituent's determination is one of the widest used application of the Near-infrared spectroscopy. Samples are exposed to electromagnetic radiation and its subsequent absorption at the wavelength of 800 to 2500 nm range is the basic principle of this technique [38].

The strongest stimulus in the development of NIR spectroscopy applications in both pharmaceutical and medical fields was due to the recent development of new softwares and sophisticated equipments. The manufacturers experienced in this fields develops advanced equipment on the regular basis [39].

NIR spectroscopy is a well-established branch of spectroscopy that measures chemical bonds on the basis of overtones and combination bands of specific functional groups [40]. Coverage of electromagnetic radiations in the range of 750 to 2500 nm is widely accepted fact of this technique [41].

Bazar G *et al.* applied principal component regression (PCR) and partial least squares regression (PLSR)for calibration on dry matter and purity [pure honey (*Robinia* honeys) content against high-fructose corn syrup (HFCS)] of pure and adulterated honeys. Spectral regions were divided into two parts; 1300–1800 and 1600–1800 nm. Former region represents the first overtone region of O–H bands (water stretching vibrations) and latter predominantly representing the first overtone region of C–H bands (carbohydrates). Adulteration of *Robinia* honeys with HFCS caused gradual decrease of structures of water molecules and increase in the interaction with other molecules. In this study, anopposing correlation was found between the dry matter content and the adulteration level of *Robinia* honeys mixed with HFCS [42].

A near infrared spectroscopy technique with Chemometrics for the detection of adulteration by jiggery in honey was performed by Chellakutty & Aravamudhan (2015) by using principal component analysis (PCA) and partial least square (PLS) techniques. The absorption data of pure and adulterated honeys were clipped in the wavelength range of 1100–2200 nm. Peak wavelengths adulterated honeys were found to be shifted compared to pure honey. Researchers concluded that suchtechniques can be useful in monitoring quality of honey between batches by estimating its contents quantity with the help of chemical analysis [43].

Rheology

The term 'Rheology' was invented by Professor Bingham of Lafayette college, Easton, Pennsylvania. It means the study of the deformation and flow of matter [44].

Foods are fluids if they take the shape of the container in which they are kept and do not retain their shape. They exhibit non-Newtonian behavior because they may contain amounts of dissolved or suspended solids [45].

The design of equipments for transportation, pumping, processing, quality control, sensory analysis and storage of honey requires its rheological knowledge. Moisture content as well as chemical composition depends upon its rheological characteristics [46].

A study conducted by Kamboj and Mishra (2015) [47] used rotational rheometer with parallel plate geometry for prediction of jaggery syrup adulterations with different honey samples. This is well evident that honey samples behave as non Newtonian fluid after adulteration. Researchers found linear relationship between concentration of adulterant (Jaggery syrup) and viscosity of the honey. An oscillatory test was used to understand the correlation between storage time on different honey samples. Researcher further found decrease in self life with increase in adulteration.

Sensors

A recent research [48] focused on evaluating the performance of sensors [Electronic nose (EN) and Electronic Tongue (ET)] and spectra [Near Infrared (NIR) and Mid Infrared (MIR)] in botanical origin classification and adulterant determination of raw honey was published by Gan Z *et al*. They found difference in performance and accuracy of sensors and analytical techniques. For NIR and MIR samples were scanned in the region 10000-4000 cm^{-1} and 4000-650 cm^{-1} respectively. For classification of botanical origin of honeyPartial least squares discriminant analysis (PLSDA) support vector machine (SVM) algorithms model and Interval partial least squares (iPLS) model were used. They found that SVM and iPLS improved accuracy of spectral analysis and sensors as compared to PLSDA. Researchers finally concluded with the result that EN is more suitable for testing freshness of honey because it's sensitive to aroma substances. For amino acids, minerals, phenols, monosaccharides and disaccharides. ET is more sensitive and for real time and onsite classification of botanical origin of honey spectral analysis is more suitable. Finally, researchers concluded that ET is more suitable for detecting adulteration in honey.

CONCLUSION

Honey is one of the most nutritious and flavored gift of the nature. This chapter starts with the discussion about various uses and composition of honey. Honey may be different types and of different floral origin. This is the reason of diversity of its composition and varieties. Different kind of analytical techniques based on Liquid Chromatography (LC), Nuclear Magnetic Resonance (NMR), Near Infrared Spectroscopy (NIS), Rheology and Sensors are discussed here in this chapter. The recent developed methods for the adulteration analysis of honey in last three years are discussed here.

CONSENT FOR PUBLICATION

Not applicable.

CONFLICT OF INTEREST

The author (editor) declares no conflict of interest, financial or otherwise.

ACKNOWLEDGEMENTS

Author acknowledges support of Dr. Brijesh Sharma for improving level of English in this paper.

REFERENCES

[1]   Manning L, Soon JM. Developing systems to control food adulteration. Food Policy 2014; 49: 23-32.
        [http://dx.doi.org/10.1016/j.foodpol.2014.06.005]

[2]   Accum FC. Available from: http://www.rsc.org/images/cw01_accum_tcm18-196603.pdf

[3]   Rasul CH. Alarming situation of food adulteration. Bangladesh Medical Journal, 2013; 46(1- 2):1-2.

[4]   Manning L, Soon JM. Food Safety, Food Fraud, and Food Defense: A Fast Evolving Literature. J Food Sci 2016; 81(4): R823-34.
        [http://dx.doi.org/10.1111/1750-3841.13256] [PMID: 26934423]

[5]   Deelstra H, Thorburn Burns D, Walker MJ. The adulteration of food, lessons from the past, with reference to butter, margarine and fraud. Eur Food Res Technol 2014; 239: 725-44.
        [http://dx.doi.org/10.1007/s00217-014-2274-0]

[6]   Khapre MP, Mudey A, Chaudhary S, Wagh V, Dawale A. Buying Practices and Prevalence of Adulteration in Selected Food Items in a Rural Area of Wardha District: A Cross - Sectional Study. Online J Health Allied Sci 2011; 10(3): 4.

[7]   Roshan AA, Gad HA, El-Ahmady SH, Abou-Shoer MI, Khanbash MS, Al-Azizi MM. Characterization and Discrimination of the Floral Origin of Sidr Honey by Physicochemical Data Combined with Multivariate Analysis. Food Anal Methods 2016; 1-10.
        [http://dx.doi.org/10.1007/s12161-016-0563-x]

[8]   Verzera A, Condurso C. Sampling techniques for the determination of the volatile fraction of honey. Comprehensive Sampling and Sample Preparation 2012; 4: 87-116.
        [http://dx.doi.org/10.1016/B978-0-12-381373-2.00129-0]

[9]   Mehryara L, Esmaiili M. Honey and honey adulteration detection: A review http://www.icef11.org/content/papers/nfp/NFP1066.pdf

[10]   Ndife J, Abioye L, Dandago M. Quality assessment of nigerian honey sourced from different floral locations. Nigerian Food J 2014; 32(2): 48-55.
        [http://dx.doi.org/10.1016/S0189-7241(15)30117-X]

[11]   Lakshmi V. Food adulteration. International Journal of Science Inventions Today 2012; 1(2): 106-13.

[12]   Fairchild GF, Capps O, Nichols JP. Estimated Impacts of Economic Adulteration on the U.S. Honey Industry. Western Agricultural Economics Association Annual Meetings, Vancouver, British Columbia, June 29 - July 1, 2000.

[13]   Eteraf-Oskouei T, Najafi M. Traditional and modern uses of natural honey in human diseases: a review. Iran J Basic Med Sci 2013; 16(6): 731-42.
        [PMID: 23997898]

[14]   Akbari B, Gharanfoli F, Khayyat MH, Khashyarmanesh Z, Rezaee R, Karimi G. Determination of heavy metals in different honey brands from Iranian markets. Food Addit Contam Part B Surveill 2012; 5(2): 105-11.
        [http://dx.doi.org/10.1080/19393210.2012.664173] [PMID: 24779739]

[15]   Mandal MD, Mandal S. Honey: its medicinal property and antibacterial activity. Asian Pac J Trop Biomed 2011; 1(2): 154-60.
[http://dx.doi.org/10.1016/S2221-1691(11)60016-6] [PMID: 23569748]

[16]   Mijanur Rahman M, Gan SH, Khalil MI. Neurological effects of honey: current and future prospects. Evid Based Complement Alternat Med 2014; 2014: 958721.
[http://dx.doi.org/10.1155/2014/958721] [PMID: 24876885]

[17]   Milojković Opsenica D, Lušić D, Tešić Ž. Modern analytical techniques in the assessment of the authenticity of Serbian honey. Arh Hig Rada Toksikol 2015; 66(4): 233-41.
[http://dx.doi.org/10.1515/aiht-2015-66-2721] [PMID: 26751854]

[18]   White JW. Advances in Food Research. Academic Press. Inc 1978; Vol. 24: pp. 288-374.

[19]   Anklam E. Evaluation of Methods of Analysis for Authenticity Proof of Honeys. 1996; pp. 1-50.

[20]   Consonni R, Cagliani LR. Recent developments in honey characterization.
[http://dx.doi.org/10.1039/C5RA05828G]

[21]   Noora MJ, Ahmad M, Ashraf MA, Zafar M, Sultana S. A review of the pollen analysis of South Asian honey to identify the bee floras of the region. Palynology 2016; 40: 1-12.
[http://dx.doi.org/10.1080/01916122.2014.988383]

[22]   da Silva PM, Gauche C, Gonzaga LV, Costa ACO, Fett R. Honey: Chemical composition, stability and authenticity. Food Chem 2016; 196: 309-23.
[http://dx.doi.org/10.1016/j.foodchem.2015.09.051] [PMID: 26593496]

[23]   Cordella C, Militão JSLT, Clément MC, Drajnudel P, Cabrol-Bass D. Detection and quantification of honey adulteration *via* direct incorporation of sugar syrups or beefeeding: preliminary study using high performance anion exchange chromatography with pulsed amperometric detection (HPAEC-PAD) and chemometrics. Anal Chim Acta 2005; 531(2): 239-48.
[http://dx.doi.org/10.1016/j.aca.2004.10.018]

[24]   Malkin AY, Isayev AI. Rheology Concepts. 2nd ed. Methods, and Applications 2012; pp. 1-8.

[25]   Yilmaz MT, Tatlisu NB, Toker OS, *et al.* Dynamic and creep rheological analysis as a novel approach to detect honey adulteration by fructose and saccharose syrups: Correlations with HPLC-RID results. Food Res Int 2014; 64: 634-46.
[http://dx.doi.org/10.1016/j.foodres.2014.07.009]

[26]   Zhou J, Qi Y, Ritho J, *et al.* Analysis of maltooligosaccharides in honey samples by ultra-performance liquid chromatography coupled with evaporative light scattering detection. Food Res Int 2014; 56: 260-5.
[http://dx.doi.org/10.1016/j.foodres.2014.01.014]

[27]   Dong H, Luo D, Xian Y, Luo H, Xin-dong G. LiC, ZhaoM. Adulteration Identification of Commercial Honeys with C-4 Sugar Content of Negative Values by Elemental Analyzer (EA) and Liquid Chromatography (LC) Coupled to an Isotope Ratio Mass Spectrometer (IRMS). J Agric Food Chem 2016; 64(16): 3258-65.
[http://dx.doi.org/10.1021/acs.jafc.6b00691] [PMID: 27064147]

[28]   Elflein L, Kurt-Peter R. Improved detection of honey adulteration by measuring differences between $^{13}C/^{12}C$ stable carbon isotope ratios of protein and sugar compounds with a combination of elemental analyzer - isotope ratio mass spectrometry and liquid chromatography - isotope ratio mass spectrometry ($\delta^{13}C$-EA/LC-IRMS). Apidologie (Celle) 2008; 39: 574-87.
[http://dx.doi.org/10.1051/apido:2008042]

[29]   Godin JP, McCullagh JS. Review: Current applications and challenges for liquid chromatography coupled to isotope ratio mass spectrometry (LC/IRMS). Rapid Commun Mass Spectrom 2011; 25(20): 3019-28.
[http://dx.doi.org/10.1002/rcm.5167] [PMID: 21953956]

[30] Wang S, Guo Q, Wang L, *et al.* Detection of honey adulteration with starch syrup by high performance liquid chromatography. Food Chem 2015; 172(172): 669-74.
[http://dx.doi.org/10.1016/j.foodchem.2014.09.044] [PMID: 25442605]

[31] Du B, Wu L, Xue X, *et al.* Rapid Screening of Multiclass Syrup Adulterants in Honey by Ultrahigh-Performance Liquid Chromatography/Quadrupole Time of Flight Mass Spectrometry. J Agric Food Chem 2015; 63(29): 6614-23.
[http://dx.doi.org/10.1021/acs.jafc.5b01410] [PMID: 26151590]

[32] Laghi L, Picone G, Capozzi F. Nuclear magnetic resonance for foodomics beyond food analysis. Trends Anal. Chem. TrAC 2014; 59: 93-102.
[http://dx.doi.org/10.1016/j.trac.2014.04.009]

[33] Mannina L, Serge A. High Resolution Nuclear Magnetic Resonance: From Chemical Structure to Food Authenticity. Grasas Aceites 2002; 53(1): 22-33.
[http://dx.doi.org/10.3989/gya.2002.v53.i1.287]

[34] Marcone MF, Wang S, Albabish W, Nie S, Somnarain D, Hill A. Diverse food-based applications of nuclear magnetic resonance (NMR) technology. Food Res Int 2013; 51(2): 729-47.
[http://dx.doi.org/10.1016/j.foodres.2012.12.046]

[35] Hoa D. HoaD. Relaxation and its characteristics: T1 and T2 times. https://www.imaios.com/en/e-Courses/e-MRI/NMR/Relaxation-nmr

[36] Ribeiro ROR, Mársico ET, Carneiro CS, Monteiro MLG, Júnior CC, Jesus EFO. Detection of honey adulteration of high fructose corn syrup by Low Field Nuclear Magnetic Resonance (LF ^{1}H NMR). J Food Eng 2014; 135: 39-43.
[http://dx.doi.org/10.1016/j.jfoodeng.2014.03.009]

[37] Spiteri M, Jamin E, Thomas F, *et al.* Fast and global authenticity screening of honey using ^{1}H-NMR profiling. Food Chem 2015; 189(15): 60-6.
[http://dx.doi.org/10.1016/j.foodchem.2014.11.099] [PMID: 26190601]

[38] Kumaravelu C, Gopal A. Detection and Quantification of Adulteration in Honey through Near Infrared Spectroscopy. Int J Food Prop 2015; 18(9): 1930-5.
[http://dx.doi.org/10.1080/10942912.2014.919320]

[39] Ciurczak EW, Igne B. Pharmaceutical and Medical Applications of Near-InfraredSpectroscopy. 2nd ed. Taylor & Francis Group 2015; p. 17.

[40] Rodriguez-SaonaLE. GiustiMM, ShottsM Advances in Infrared Spectroscopy for Food Authenticity Testing Advances in Food Authenticity Testing. Elsevier Ltd. 2016; p. 74.

[41] Lin M, Rasco BA, Cavinato AG. Al-HolyM Infrared (IR) Spectroscopy—Near Infrared Spectroscopy and Mid-Infrared Spectroscopy Infrared Spectroscopy for Food Quality Analysis and Control. Elsevier Inc 2009; p. 120.

[42] Bázár G, Romvári R, Szabó A, Somogyi T, Éles V, Tsenkova R. NIR detection of honey adulteration reveals differences in water spectral pattern. Food Chem 2016; 194: 873-80.
[http://dx.doi.org/10.1016/j.foodchem.2015.08.092] [PMID: 26471630]

[43] Abbas O, Dardenne P, Baeten V. Near-Infrared, Mid-Infrared, and Raman Spectroscopy. Chemical Analysis of Food: Techniques and Applications. Elsevier, 2012, pp 2 Inc..

[44] Taghi M. Viscosity and Oscillatory Rheology.Practical Food Rheology An Interpretive Approach. Blackwell Publishing Ltd 2011; p. 8.

[45] Nayik GA, Dar BN, Nanda V. Physico-chemical, rheological and sugar profile of different unifloral honeys from Kashmir valley of India. Arab J Chem 2015.
[http://dx.doi.org/10.1016/j.arabjc.2015.08.017]

[46] Kamboj U, Mishra S. Prediction of adulteration in honey using Rheological parameters. Int J Food Prop 2015; 18(9): 2056-63.

[http://dx.doi.org/10.1080/10942912.2014.962656]

[47]   Gallegos C, Walters K. Rheology, Encyclopedia of Life Supporting Systems. Eolss Publishers Co. Ltd. 2010; Vol. I: p. 2.

[48]   Gan Z, Yang Y, Li J, *et al.* Using Sensor and Spectral Analysis to Classify Botanical Origin and Determine Adulteration of Raw Honey. J Food Eng 2016; 178: 151-8. [http://dx.doi.org/10.1016/j.jfoodeng.2016.01.016]

CHAPTER 5

UV Spectrophotometry: Applications in Adulteration Analysis of Drugs

Alankar Shrivastava[*]

Department of Pharmaceutical Quality Assurance, Lloyd School of Pharmacy, Greater Noida, Knowledge Park II, Greater Noida, Uttar Pradesh 201310, India

Abstract: UV Vis spectrophotometry is one of the basic analytical techniques used for the analysis of many inorganic and organic molecules. It is also used for the determination of complex mixtures to some extent. UV Vis spectrophotometry is used for both industrial as well as academic researches. This chapter includes determination in adulteration of some drugs by using this versatile technique. This chapter provides description of research related to the adulteration analysis of drugs by UV spectrophotometry. The chapter also contains a brief discussion about this technique initially followed by published applications in the field of adulteration analysis.

Keywords: Adulteration, *Aloe vera*, Aloin, Aphrodisiacs, Derivative spectrum, Lambert-beer law, Spectrophotometry, Spectrum, Street drugs, Ultra Violet, Viagra, Visible, Wavelength.

INTRODUCTION

UV-Visible spectroscopy is one of the most basic techniques for testing a variety of materials. Different analytes *e.g.* conjugated organic compounds, transition metal complexes and biological macromolecules etc. can be determined both qualitatively and quantitatively [1].

UV-visible spectroscopy is widely used for the analysis of chromophores (groups of atoms characterized by strongly absorbing electronic transitions). Direct relation of the spectra to molecular functional groups and simplicity of spectra are the attractiveness of this method [2].

UV/Visible spectrophotometers are widely used by many laboratories, including in academics, research as well as in industrial quality assurance [3].

---

[*] **Corresponding author Alankar Shrivastava:** Department of Pharmaceutical Quality Assurance, Lloyd School of Pharmacy, Greater Noida, Knowledge Park II, Greater Noida, Uttar Pradesh 201310, India; Tel: +91-7351002560; E-mail: alankarshrivastava@gmail.com

The color of light depends upon its wavelength. The perception of red is produced by the longest wavelength, while violet is produced by the shortest ones. Thus each color has a different wavelength. The green light, for example, is produced between wavelength 520–565 nm [4].

A definition of color [5] might be [as published by Ohta & Robertson (2005)]:

'A visual perception that enables one to differentiate otherwise identical objects by the intensity and the wavelength of light'.

UV-Vis Spectra

The wavelength of electromagnetic radiations between 190 nm to 800 nm is used in UV Vis Spectroscopy. This region is further divided into two parts *i.e.* UV, 190-400 nm, and VIS, 400-800 nm. Since the changes in the electronic energy of the molecule are due to absorption of ultraviolet and visible regions, it is often called as electronic spectroscopy [6].

After receiving electromagnetic radiation, molecules absorb energy and undergo translational, rotational or vibrational motion or ionization depending upon the frequency. The unstable excited molecules then drop down to the ground state again and release energy. This absorption or emittance of energy is recorded to get a spectrum which gives information about the matter under study [7].

Light consists of perpendicular oscillating magnetic and electric fields whose energy is given by:

$$E = hc/\lambda = h\nu$$

where, 'h' is the Planck's constant, c is the speed, 'λ' is the wavelength, and 'c' is the frequency of the light. When the energy of the targeted molecule starts receiving the amount of energy between ground and excited state, absorption occurs. This absorbed energy from the photons of light will be released in the form of radiations or heat [8].

Molecules which absorb photons of energy corresponding to wavelengths in the range 190 mm to about 1000 nm exhibit UV/Visibleabsorption spectra. The quantized internal energy E_{total} of a molecule in its electronic ground or excited state can be mathematically written with sufficient accuracy for analytical purposes, by

$$E_{total} = E_A + E_B + E_C$$

where E_A is the electronic, E_B the vibrational and E_C the rotational energy, respectively.

There is a change in rotational and vibrational energies by absorption of photon. Vibronic transition (electronic + vibrational transition) corresponding to an absorption band consists of rotational lines. No rotational structure is distinguishable in the case of liquids and solids because the rational lines are broad and overlap [9].

The wavelength ranges of $4.0 - 7.0 \times 10^{-7}$ m or 400-700 nm is of visible light range. After the absorption of light, the outer valence electrons are promoted from ground state (least energy state) to higher energy (excited) state. The frequency of the light is directly related to the energy of visible light, *e.g.* 170 kJ mol^{-1} and 300 kJ (moles of photon) for red and blue light, respectively. The transitions of electrons from ground state to higher energy state can also occur in ultraviolet region [10]. The wavelength and the direction of propagation is explained in Fig. (1).

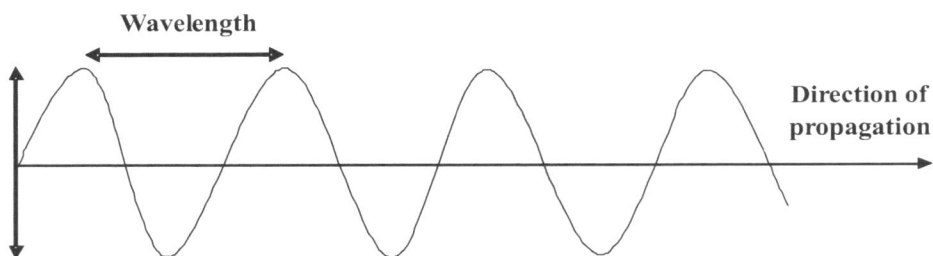

Fig. (1). Wave of electromagnetic radiation.

Lambert-Beer law

There is an absorption of radiation when light passes through the material. This law relates to this property of the material. The equation forms of this law is:

$$\log T = \log(I/I_o) = \varepsilon l c$$

where 'T' is the transmittance, 'I' is the incident light intensity, 'I_0' is the transmitted light intensity, 'ε' is the absorption coefficient, 'l' is the distance travel by the radiation through the material or path length.

The negative logarithm of transmittance equal to absorbance; thus absorbance can be mathematically represented as:

$$A = -\log T$$

And finally the law got its shape in the form of following equation:

$$A = \varepsilon l c$$

There are three components of a typical UV-Vis spectrometer, in addition to the mechanism for holding the sample in place. Either diffraction grating or monochromator used as a source of light. The detector can either be a photodiode (used with a monochromator) or a charge coupled device (CCD, used with diffraction gratings). The third component is the radiation source and is generally a tungsten filament, or advanced versions like deuterium arc lamps and xenon arc lamps [11].

DERIVATIVE SPECTROPHOTOMETRY

Derivative spectrophotometry has acquired increasing importance especially accepted in UV/VIS spectrophotometry and is now undergoing vigorous development. Unlike gases and vapors, which frequently give many sharp lines in the UV and in the visible part of the spectrum, dissolved substances or liquids usually give less-characteristic spectra with more or less developed maxima and shoulders, which must be ascribed to the overlapping of several terms of electron transitions, such as n→π*, σ→σ*, and π→π* and to rotation and translation. Derivative spectra make it possible to determine flat maxima more precisely, to isolate shoulders and weak signals from an unwanted background [12].

Quantitative analysis of multi-component mixtures is the application of derivative spectrophotometry. The derivative spectra are more structured, thus identify very tiny differences when compared with the original spectra. This method is widely used to improve signal and resolve overlapped peaks because it differentiates between closely adjacent peaks and weak peaks identification that are hidden by sharper peaks [13]. As compared to zero order spectra, derivative spectra are always more complex [14].

Analysis of Viagra (Sildenafil) and Cialis (Tadalafil) Tablets

In a study, Obeidat & Al-Tayyem (2014) investigated adulteration of Sildenafil and Tadalafil tabletsbyUV spectrophotometric techniquescoupled with chemometrical algorithms such as Principal Components Analysis (PCA) and Cluster Analysis (CA). Authors also used Near-infrared (NIR) and compared the two methods. They concluded that both portable NIR and the UV-Visible techniques coupled with chemometricsgive good and consistent results in discriminating among the original and the manipulated drug. Portable NIR is a non-destructive method that needs no sample pretreatment but expensive technique. On the other hand, the UV-Visible technique is less expensive but needs sample pretreatment which is not always easy [15].

Analysis of Aloe Barbadensis

Aloe barbadensis Miller, also known as *Aloe vera*, belongs to the Liliaceae family. More than 400 species belong to this family [16]. There are about 250 different species of aloe found worldwide [17] but there are only a few having commercial importance. For medicinal properties, *A. barbadensis* Miller (popularly called *Aloe vera*) and *Aloe aborescens*are grown commercially. *Aloe vera*is an important component of traditional medicines of many ancient cultures such as India, China, Japan and West Indies. It has a history of usage in folk medicines for over 2000 years. *A. vera*is considered to be the most potent and popular plant in the pharmaceutical research field [16]. The *A. vera* plant matures within 4 years and its total lifespan is about 12 years. The *Aloe vera* gel is referred to be a jelly obtained from the leaves of inner parenchymal cells and is mucilageneous in nature [17].

Some biological properties attributed to the gel of this plant are in the form of its moisturizing effect, antibacterial, antiviral, laxative, anti-inflammatory, protection from radiation and immunostimulation [18]. The active components are aloin, aloesin, aloemannan, aloeride, aloemodin, naftoquinones, flavonoids, methychromones, acemannan, saponin, aminoacid, sterols and vitamins [19].

Aloe species were extensively studied by researchers worldwide for isolations of active principles and their biological activities. Aloin (Fig. **2**) and aloe-emodin are the main active components identified in the plant. Aloin showed antioxidant or pro-oxidant effect on plasmid DNA, depending on their nature (structure) and their concentrations. Its pro-oxidant or antioxidant effect on DNA may be due to the balance of two activities, free radical-scavenging activity and reducing power on iron ions, which may drive the Fenton reaction *via* reduction of iron ions [20].

Sharma *et al.* [21] authenticated *Aloe vera* samples by estimating Aloin content by using UV spectrophotometry. The concentration ranges of standard Aloin solution prepared was 20-100μg/ml and absorbance was measured at 357 nm taking methanol as blank. The UV study was also carried out to determine the amount of Aloin in all the samples and to detect the presence of matrix and its effects. The results show practically an interference of matrix component over the response factor of target analyte; Aloin. They found formulation in which whole aloe used will contain more Aloin content as compared to inner gel formulations. They concluded that out of 4 samples, only one sample shows the presence of Aloin content and that is also in very low amount.

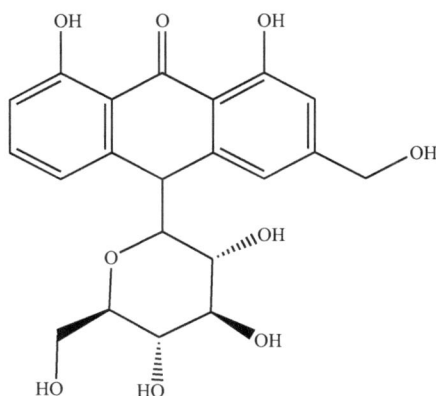

Fig. (2). Aloin, 10-glucopryanosyl-1,8-dihydroxy-3-(hydroxymethyl)-9(10H)-anthracenone.

Adulteration Analysis in Herbal Aphrodisiac

To reduce the symptoms or to treat or prevent chronic diseases, the usage of botanical and herbal medicines is on the rise [22]. The western population preferably opts for natural remedies for safe and effective treatment after considering the adverse effects of the synthetic drugs. For primary healthcare, it is well documented that 80% of the total world population has faith in medicines used traditionally, particularly including plant drugs [23].

On the name 'Aphrodite' of Greek goddess of love, sex and beauty, the word 'Aphrodisiac' is derived. The definition of Aphrodisiacs is:

An aphrodisiac is defined as an agent (food or drug) that arouses sexual desire [24].

About 10-52% of men and 25-63% of women reported to have sexual dysfunction (ED) and a serious medical and social symptom. ED is defined as:

The consistent inability to obtain or maintain an erection for satisfactory sexual relations [25].

Sildenafil is an orally active, potent and selective cyclic guanosine monophosphate (cGMP) specific phosphodiesterase type 5 (PDE5) inhibitor [26]. The scientists of Pfizer, USA discovered this drug and on 27th March 1998 received approval from the US FDA [27]. The initial dose of sildenafil is 50 mg (nmt once/day) taken 1 hr before sexual activity. It absorbs well during fasting and reaches maximum plasma concentrations within 30-120 min. Its half-life is 4 hr and excreted by the liver [28].

Pandey & Parikh (2015) [29] detected the presence of sildenafil citrate in aphrodisiac herbal formulations. For extraction, a mixture of methanol: water (50:50) was used. Sildenafil citrate is soluble in methanol: water solution and shows λ_{max} at 292 nm. Any absorbance at this wavelength confirms the adulteration of Sildenafil. Researchers scanned the extracts of various herbal aphrodisiac formulations in the range of 200-400 nm to detect any absorbance at 292 nm. Researchers detected two out of seven products containing Sildenafil which are claimed to be 'All natural' herbal supplements. Thus, UV spectrophotometry finds novel application of detection of Sildenafil in such kinds of formulations.

Analysis of Antidiabetic Tablet

Diabetes mellitus is a serious and common metabolic disorder with multiple complications and diverse causes [30]. Diabetes mellitus is affecting about 5-10% population of adults of western world and is one of the common chronic disorders [31]. According to the estimation of International Diabetes Federation (IDF), this disorder is expected to increase to 522 million by 2030 as compared to 366 million in 2011 [32].

Diabetes mellitus is characterized by disturbances of metabolism of carbohydrate, fat and proteins with elevated blood glucose levels [33]. The modern approach to control or treat diabetes such as insulin and oral synthetic antidiabetic drugs is producing side effects and unsatisfactory results [34]. This is the reason of inclination of people towards more safe and effective herbal antidiabetic medications. But in the next report, such kind of antidiabetic formulation was found adulterated with synthetic drug.

Kumar M with coworkers [35] identified the major ingredient in the herbal anti diabetic pills as metformin hydrochloride. They used UV spectrophotometry method for the quantitation and found 93.1 mg of Metformin base per 100 mg of the pill. Researchers also demonstrated dissolution profile of the pill and found to be unmatched with the pharmacopoeial requirement for release of any metformin tablets.

Analysis of Street Drug Powders

A common perception about street drugs such as heroin and ecstasy is that they are adulterated to dilute the product during distribution to increase the profit [36]. Thus, this provides an excellent opportunity of profit for dealers and smugglers in the illegal market. Several different adulterants have been found, such as salts, pharmaceuticals, solvents, sugars, ashes, portions of dry fruits, glass, plastics and silica [37]. The pharmacological effect of the drug can be altered by adding any biologically active drug in an unpredictable way [38].

A spectrophotometric method for the resolution of mixtures of cocaine, procaine and lidocaine in street drug powder samples was proposed by Cruz *et al.* [39]. The order of determinations was procaine, cocaine and lidocaine. Phosphate buffer pH 7.8 was used for the analysis. The reason is to provide acceptable peak resolution and signal intensity, in addition to good reproducibility. This method allows both adulterants to be quantified and hence the drug purity to be assessed.

The sequential technique involves deconvoluting a mixed spectrum into the individual spectra of the mixture components, which are determined one by one, whereas inconventional spectrometric methods, all components are determined on the basis of the same spectrum simultaneously. A second derivative spectrum was used for the determination between 380-220 nm. The procaine, cocaine and Lidocaine were quantified by the spectral difference between 321-370, 291-370, 277-271 nm respectively. Authors concluded that these kinds of spectral libraries are useful for the determination of adulteration in such illicit drugs.

Detection of Adulterated Ginkgo Biloba Supplements

Since *Ginkgo biloba* L is the single extinct species of the family Ginkgoaceae, it is also known as living fossil [40]. This is believed to be one of the oldest living tree of about more than 250 million years ago [41].

The extracts have many therapeutic applications and have broad spectrum pharmacological activities. These include cognitive impairment, liver and kidney injury, depression, cerebrovascular insufficiency, peripheral arterial occlusive disease and myocardial ischemia [42].

Hamly JM *et al.* [43] utilized UV spectrophotometry for acquiring fingerprints of 18 commercially available Ginkgo biloba supplements, 12 samples of raw *G. biloba* leaves and three *G. biloba* standard reference materials from the National Institute of Standards and Technology. They found 3 commercial products adulterated with rutin, four with quercetin, and one sample with unidentified flavanoid. Samples were also analyzed by HPLC and NIR spectroscopy. The NIR spectrophotometry was found to be unsuitable in this case because of the presence of excipients. One-class soft independent modeling of class analogy (SIMCA) and Principle Component Analysis (PCA) were used for data analysis.

CONCLUSION

However, the number of studies published by UV spectrophotometry in the field of adulteration analysis are few but opens a new application part of this technique.

To the best of knowledge of authors, this kind of literature is not published anywhere focusing specifically in this part. Adulteration analysis of Sildenafil and Tadalafil tablets, *Aloe barbadensis*, herbal Aphrodisiac, antidiabetic herbal pills, street powder drugs and *Ginkgo biloba* Supplements are some of the documented application of UV spectrophotometry.

UV Vis technique is one of the oldest and classic technique for the analytical determinations in both pharmaceutical industry and academic research. This chapter described the application of UV Vis Spectrophotometry in the determination analysis of some drugs. However, limited applications found may be because of more advance technologies today.

CONSENT FOR PUBLICATION

Not applicable.

CONFLICT OF INTEREST

The author (editor) declares no conflict of interest, financial or otherwise.

ACKNOWLEDGEMENT

Author acknowledges support of Dr. Brijesh Sharma for improving level of English in this paper.

REFERENCES

[1]   UV-Visible Analysis of Bitterness and Total Carbohydrates in Beer. 2013. https://www.thermo scientific.com/content/dam/tfs/ATG/CAD/CAD%20Documents/Application%20&%20Technical%20 Notes/Molecular%20Spectroscopy/UV%20Visible%20Spectrophotometers/Spectrophotometer%20Sy stems/AN52467-E-1113M-UVVisBeer-H.pdf

[2]   Doyle WM, Tran L. Analysis of strongly absorbing chromophores by UV-visible ATR spectroscopy. Spectroscopy (Springf) 1999; 14(4): 46-54.

[3]   Steve Upstone. 2012. http://www.perkinelmer.com/CMSResources/Images/44-136839TCH_Vali dating_UV_Visible.pdf

[4]   Malacara D. Color vision and colorimetry: theory and applications. 2nd ed. Bellingham: SPIE Press 2011; p. 1.

[5]   Noboru O, Robertson AR. Wiley–IS&T Series in Imaging Science and Technology. England 2005; p. xv.

[6]   Kumar S. Spectroscopy of Organic Compounds. pp. 4. Available from: http://www.uobabylon.edu.iq/ eprints/publication_11_8282_250.pdf

[7]   Hamid H. Ultraviolet and Visible Spectrophotometry http://nsdl.niscair.res.in/jspui/bitstream/12345 6789/772/1/revised%20Ultraviolet%20and%20Visible%20Spectrophotometry.pdf

[8]   Hemyk M, Volkin DB, Burke CJ, Middaugh CR. Ultraviolet Absorption Spectroscopy. Protein Stability and Folding, Volume 40, Methods in Molecular Biology. 1995. Humana Press. pp. 92.

[9]   Laqua K, Melhuish WH, Zander M. Molecular Absorption Spectroscopy, Ultraviolet And Visible

(UV/VIS). Pure & App/. Chem 1988; 60(9): 1449-60.

[10]   Ultraviolet/visible spectroscopy. The Royal Society of Chemistry. Available from: http://media.rsc.org/Modern%20chemical%20techniques/MCT4%20UV%20and%20visible%20spec.pdf

[11]   Lindsay H. UV-Vis Spectrometers. R & D Magazine, 2010. Available from: http://www.rdmag.com/sites/rdmag.com/files/legacyfiles/RD/Tools_And_Technology/2010/04/RD04 UV-Vis_spec.pdf

[12]   Talsky G, Mayring L, Kreuzer H. High-resolution, higher-order UV VIS –derivative spectrophotometry. Angew Chem 1978; 17: 785-99. a
[http://dx.doi.org/10.1002/anie.197807853]

[13]   Ojeda CB, Rojas FS. Recent applications in derivative ultraviolet/visible absorption spectrophotometry: 2009–2011. Microchem J 2013; 106: 1-16.
[http://dx.doi.org/10.1016/j.microc.2012.05.012]

[14]   Rojas SF, Ojeda CB. Recent developments in derivative ultraviolet/visible absorption spectrophotometry. Anal Chim Acta 2004; 518: 1-24.
[http://dx.doi.org/10.1016/j.aca.2004.05.036]

[15]   Obeidat SM, Al-Tayyem B. Uncovering Counterfeit Viagra and Cialis Using Portable NIR and UV-Visible Spectroscopy Coupled with Multivariate Data Analysis (a preliminary study). Jordan J Chem 2014; 9(3): 159-69.
[http://dx.doi.org/10.12816/0026398]

[16]   Radha MH, Laxmipriya NP. Evaluation of biological properties and clinical effectiveness of Aloe vera: A systematic review. J Tradit Complement Med 2014; 5(1): 21-6.
[http://dx.doi.org/10.1016/j.jtcme.2014.10.006] [PMID: 26151005]

[17]   Javed S. Atta-ur-Rahman. Aloe Vera Gel in Food, Health Products, and Cosmetics Industry. Edited by Atta-ur-Rahman. Studies in Natural Products Chemistry, Vol. 41. 2014 Elsevier. 261-285.

[18]   Femenia A. High-value co-products from plant foods: cosmetics and pharmaceuticals. Editor: Waldron K. Handbook of waste management and co-product recovery in food processing. Volume 1, Woodhead Publishing Limited, Cambridge, 2007, p. 475.
[http://dx.doi.org/10.1201/9781439824498.ch18]

[19]   Paez A, Gonzalez ME, Tschaplinski TJ, Tschaplinski TJ. Growth, soluble carbohydrates, and aloin concentration of Aloe vera plants exposed to three irradiance levels. Environ Exp Bot 2000; 44(2): 133-9.
[http://dx.doi.org/10.1016/S0098-8472(00)00062-9] [PMID: 10996366]

[20]   Tian B, Hua Y. Concentration-dependence of prooxidant and antioxidant effects of aloin and aloe-emodin on DNA. Food Chem 2005; 91: 413-8.
[http://dx.doi.org/10.1016/j.foodchem.2004.06.018]

[21]   Sharma V, Prajapati RPK, Shukla VJ. Detection of adulteration and authentication of Aloe Vera products by analytical and pharmacognostical tools. Intern J Pharmacy & Pharmaceutical Sci Res 2014; 4(1): 18-24.

[22]   Thomson C, Lutz RB. Herbs and Botanical Supplements: Principles and Concepts. Editors: Carol J. Boushey, Ann M. Coulston, Cheryl L. Rock, Elaine Monsen. Nutrition in the Prevention and Treatment of Disease. 2001, Academic Press, p 251.

[23]   Dubey NK, Kumar R, Tripathi P. Global promotion of herbalmedicine: India's opportunity. Curr Sci 2004; 86(1)

[24]   Singh R, Singh S, Jeyabalan G, Ali A. An overview on traditional medicinal plants as aphrodisiac agent. J Pharmacognosy&Phytochemistry 2012; 1(4): 43-56.

[25]   Kotta S, Ansari SH, Ali J. Exploring scientifically proven herbal aphrodisiacs. Pharmacogn Rev 2013;

7(13): 1-10.
[http://dx.doi.org/10.4103/0973-7847.112832] [PMID: 23922450]

[26]   Trussell JC, Anastasiadis AG, Padma-Nathan H, Shabsigh R. Oral Type 5 Phosphodiesterase Therapy for Male and Female Sexual Dysfunction. Editors: Seftel AD, Padma-Nathan H, McMahon CG, Giuliano F, Althof SE, Male and Female sexual dysfunctions. 2004, Elsevier, p. 108.

[27]   Badwan AA, Nabulsi L, Al-Omari MM, Daraghmeh N, Ashour M. Sildenafil Citrate Analytical Profiles of Drug Substances and Excipients. Academic Press 2001; Vol. 27: pp. 339-76.

[28]   Tariq SH, Morley JE. Erectile Dysfunction Encyclopedia of Neuroscience. Elsevier Ltd 2009; p. 1185.

[29]   Pandey A, Parikh P. Detection of sildenafil citrate from aphrodisiac herbal formulations. Int J Pharm Sci Res 2015; 6(9): 4080-5.
[http://dx.doi.org/10.13040/IJPSR.0975-8232.6(9).4080-85]

[30]   McCowen KC, Smith RJ. Diabetes Mellitus, Classification and Chemical Pathology. 2013 Elsevier Ltd. Encyclopedia of Human Nutrition, Volume 2, p. 12.

[31]   Raffel LJ, Goodarzi MO. Diabetes Mellitus Reference Module in Biomedical Research. Elsevier Inc 2014; p. 1.

[32]   Alam U, Asghar O, Azmi S, Malik RA. General aspects of diabetes mellitus. Handbook of Clinical Neurology, Vol. 126 (3rd series) Diabetes and the Nervous System D.W. Zochodne and R.A. Malik, Editors. 2014 Elsevier, p. 211.
[http://dx.doi.org/10.1016/B978-0-444-53480-4.00015-1]

[33]   Sonia TA, Sharma CP. Diabetes mellitus – an overview. Oral Delivery of Insulin Elsevier Limited 2014; p. 1.

[34]   Peng-Cheng W, Shan Z, Bing-You Y, Qiu-Hong W, Hai-Xue K. Hepatoprotective and antidiabetic effects of *Pistacialentiscus* leaf and fruit extracts. Carbohydr Polym 2016; 148(5): 86-97.
[http://dx.doi.org/10.1016/j.carbpol.2016.02.060] [PMID: 27185119]

[35]   Kumar M, Mandal V, Hemalatha S. Detection of metformin hydrochloride in a traditionally used indian herbal drug for antidiabetic: a case report. Int J Pharma Bio Sci 2011; 2(2): 307-13.

[36]   Coomber R, Measham F, McElrath K, Moore K. Key Concepts in Drugs and Society. London: Sage 2013; p. 161.
[http://dx.doi.org/10.4135/9781526401670]

[37]   Orlando N, Bruno N. Adulterants Found in Mixtures of Illegal Psychoactive Drugs http://bdigital.ufp.pt/bitstream/10284/938/1/208-218.pdf

[38]   Schneider S, Meys F. Analysis of illicit cocaine and heroin samples seized in Luxembourg from 2005-2010. Forensic Sci Int 2011; 212(1-3): 242-6.
[http://dx.doi.org/10.1016/j.forsciint.2011.06.027] [PMID: 21767923]

[39]   Chukwunonso Obi B. ChinwubaOkoye T, Okpashi VE, NonyeIgwe C, OlisahAlumanah E. Comparative Study of the Antioxidant Effects of Metformin, Glibenclamide, and Repaglinide inAlloxan-Induced Diabetic Rats. J Diabetes Res 2016; 2016: 1635361.
[http://dx.doi.org/10.1155/2016/1635361] [PMID: 26824037]

[40]   Soo L, Kyong SP. The Use of Ginkgo biloba Extract in Cardiovascular Protection in Patients with Diabetes. Diabetes: Oxidative Stress and Dietary Antioxidants. Elsevier Inc, p. 165.

[41]   Sasaki K, Wada K, Haga M. Chemistry and biological actvities of ginkgo biloba. Atta-ur-Rahman (Ed.) Studies in Natural Products Chemistry, Vol. 28, 2003 Elsevier Science, p. 165.

[42]   Margitta D, Robert WC. Ginkgo biloba. Nutraceuticals. 2016, Elsevier Inc, p. 681.

[43]   Harnly JM, Luthria D, Chen P. Detection of adulterated *Ginkgo biloba* supplements using chromatographic and spectral fingerprints. J AOAC Int 2012; 95(6): 1579-87.
[http://dx.doi.org/10.5740/jaoacint.12-096] [PMID: 23451372]

CHAPTER 6

Adulteration Analysis of Pomegranate Juice

Özge Taştan[*] and **Taner Baysal**

Ege University, Faculty of Engineering, Department of Food Engineering, 35100 Bornova, Izmir, Turkey

Abstract: Pomegranate juice adulteration is a common phenomenon in the market. The main reasons are associated with high product demand, high price, limited harvest season, and lack of production in some region. The most common adulteration methods are: (i) dilution with water, (ii) adding sugars or sweet juices, (iii) adding a part of lemon juice, (iv) adding fruit juices with intense red colour, and (v) adding liquids which have lower price like grape, peach or pear juice. To protect the consumer and to prevent unfair competition, authenticity and compliance with the product specification must be guaranteed. An adulterated pomegranate juice can be analysed with the determination of chemical composition. The methods commonly used for detection of adulterated pomegranate juice are the profiling and quantification of some compounds such as carbohydrates, phenolic compounds, amino acids, anthocyanins and pigments, and organic acids. The traditional chemical analysis techniques such as high-performance liquid chromatography (HPLC), gas chromatography (GC), and attenuated total reflection (ATR)-Fourier transform infrared (FTIR) spectroscopy have also been successfully performed for the detection of the authenticity of pomegranate juices. This review summarizes the adulteration methods and analysis of pomegranate juice.

Keywords: Amino acids, Adulteration, Authenticity, Chemical analysis techniques, Fourier transform infrared spectroscopy, Fourier Transform Infrared (FTIR) spectroscopy, High-performance liquid chromatography, International Multidimensional Authenticity Algorithm Specifications (IMAS), Organic acids, Phenolic compounds, Pomegranate juice, Stable isotope ratio analysis (SIRA).

INTRODUCTION

The fruit juice industry is the world's fastest-growing sector of the beverage industry. Fruit juices are necessary for the human diet, so they have often preferred due to their health benefits [1].

---

[*] **Corresponding author Özge Taştan:** Ege University, Faculty of Engineering, Department of Food Engineering, 35100 Bornova, Izmir, Turkey; Tel: +90 232 311 3043; E-mail: otastan07@gmail.com

Alankar Shrivastava (Ed.)

Pomegranate (*Punica granatum L.*) is a tree that originated in the region of Iran and cultivated since ancient times throughout the Mediterranean region and northern India. Pomegranate and its products have been used in both as traditional medicines and nutritional supplements due to beneficial effects for years. In the past decade, many types of research reported that the antioxidant activity of pomegranate juice was higher than that of the other fruit juices. Moreover, pomegranate juice consumption is recommended to be beneficial against various pathologies such as cancer, Alzheimer's disease, and heart disease. Pomegranate juice decreases cholesterol and low-density lipoprotein (LDL) levels whereas increasing high-density lipoprotein (HDL) levels, and the prostate specific antigen (PSA) [2, 3].

Significant differences were determined the chemical composition of pomegranates, depending on cultivar, growing region, maturity, environmental conditions, storage conditions and handling practices [4]. Parallel to the increase in both interest and demand, pomegranate juice adulteration has also become the primary concern of fruit juice industry [5].

History, Economy and Potential Adulterants

Nowadays, pomegranate juice adulteration has identified on account of various factors are associated with high product demand, high price, limited harvest season, and lack of production in some region. Mixing with other juices is also performed to compensate the adverse effects of low-quality raw materials and processing. Thus, some companies may intentionally add other fruit juices to make up for (i) the typical intense astringency of juice prepared with an extended maceration of the juice with the rind of fruit or peel and (ii) the pale brown color of the juice caused by the loss of anthocyanins during pasteurization [6].

The economic value of fruit juices causes the product predisposed to adulteration. It has a negative impact not only on the consumer, which expect that manufacturers and retailers provide authentic fruit juices but also in the industry, where quality authentic products have to compete with less expensive adulterated products. It also has the responsibility to comply with labelling legislation [1].

The most typical or detected adulteration methods are: (i) adding sugars or sweet juices like peach juice, in order to mask the astringency of tannins, (ii) adding a low volume of lemon juice to mask the intense sweetness of some pomegranate cultivars, (iii) adding fruit juices with dark and intense red colour, like grape or raspberry juice, (iv) adding cheap and widely available juices like grape, peach or pear juice, (v) adding anthocyanins from grape skin, elderberry, black currant, or black carrot for detection of a typical anthocyanin profile, (vi) adding cane sugar or corn sugar as detected by stable isotope ratio mass spectrometry, presence of sucrose and maltose, (vii) adding sorbitol-containing fruit juices such as apple, pear, cherry as identified by the presence of non-pomegranate anthocyanins, elevated levels of sorbitol, malic acid, or sucrose, (viii) adding citric acid as detected by low isocitric acid and high citric/isocitric acid ratios. The juices used to adulterate pomegranate juice should be readily available, cheap and with a chemical composition, color and volatile profile similar to those of pomegranate [6, 7]. An adulterated pomegranate juice can be detected if its chemical composition differs significantly from or is outside the normal range of a pure juice [6]. AIJN [8] prepared the reference guideline for fruit juices, and it is a major source of authenticity and quality control.

On the basis of extensive analytical studies on authentic juices as well as commercial samples sponsored by the Association of the German Fruit Juice Industry in collaboration with experts from research, industry and food control, the Association has formulated and published from time to time 'Richwerte und Schwankungsbreiten bestimmer Kennzahlen' or RSK for fruit juice manufactured and marketed in Germany. The meaning of RSK is guide value, range, and reference number. The guide value indicates the value which seldom falls below and rarely exceeds the specified data. The range shows the variations in the chemical composition of standard fruit juice components, deviations from which may be due to raw materials used, not appropriate additives or technical procedures. The central value is not identical with the mean, but according to the experience of all experts, it is the value of which the values of individually produced fruit juices are mostly accumulated [9]. Some authors reported that characteristic values for analytical properties of the pomegranate juice as shown in Tables **1** and **2**.

Table 1. Descriptive values for analytical properties of the pomegranate juice at 14 °Bx N=23 [10].

Analytical Properties	Range of Variation	Mean Value	Standard Error	Coefficient of Variation
Titratable acidity[a] (g/L)	8.3-17.4	13.8	±0.58	20.1
Citric acid (g/L)	6.6-13.6	11.5	±0.48	20
L-Malic acid (g/L)	0.5-0.9	0.6	±0.02	18.3
D-Isocitric acid (mg/L)	3.9-86	48.4	±4.70	47
Formol number[b]	8.6-16.2	12.9	±0.41	15.1
Glucose (g/L)	46-66	52	±1.01	9.3
Fructose (g/L)	48-70	56	±1.21	10.3
Sucrose (g/L)	0-1.5	0.2	±0.08	189.7
Total sugar (g/L)	96-137	108	±2.05	9.1

(Table 1) contd.....

Analytical Properties	Range of Variation	Mean Value	Standard Error	Coefficient of Variation
Glu/ Fru ratio	0.7-1.1	0.9	±0.02	7.9
Potassium (K) (mg/L)	2093-2517	2288	±23.70	5
Calcium (Ca) (mg/L)	11-149	52	±6.59	61
Magnesium (Mg) (mg/L)	21-104	62	±4.93	39.4
Phosphorus(P) (mg/L)	93-151	112	±2.84	12.1
Sodium (Na) (mg/L)	20-128	66	±5.90	43.1

[a]: calculated as anhydrous citric acid [b]: mL 0.1N NaOH/100 mL

Table 2. Confidence intervals (99%) and AIJN proposal for individual analytical properties of pomegranate juice [10].

Analytical properties	Confidence interval of 99%		AIJN proposal	
	min	max	min	max
Titratable acidity[a] (g/L)	12.1	15.4	10.0	15.0
Citric acid (g/L)	10.2	12.9	1.0	48.0
L-malic acid (g/L)	0.56	0.69	-	1.5
D-isocitric acid (mg/L)	35	62	10	140
Formol number[b]	11.8	14.1	5	20
Glucose (g/L)	49	55	40	80
Fructose (g/L)	53	59	45	100
Glucose/fructose	0.9	1.0	0.8	1.0
Sucrose (g/L)	0.04	0.41	-	0
Sodium (Na) (mg/L)	50	83	-	30
Potassium (K) (mg/L)	2221	2355	1300	3000
Magnesium (Mg) (mg/L)	48	76	20	110
Calcium (Ca) (mg/L)	33	70	5	120
Phosphorus(P) (mg/L)	103	119	50	170

[a]: calculated as anhydrous citric acid [b]: mL 0.1N NaOH/100 mL

Pomegranate juice adulteration can be detected five steps as shown in Table **3**.

Table 3. Detection of pomegranate juice adulteration [7].

Step of detection	Parameter
Step 1	Measure pomegranate polyphenols -Anthocyanins -Ellagitannins

(Table 3) contd.....

Step of detection	Parameter
Step 2	Measure sugar profile -Individual sugar analysis -Carbon SIRA -Brix
Step 3	Measure organic and amino acids -Individual organic acid analysis -Proline
Step 4	Measure mineral -Potassium
Step 5	Compare with IMAS 100% pomegranate juice reference

Table **4** presents the criteria for pomegranate juice authentication, a new International Multidimensional Authenticity Specifications (IMAS) algorithm was developed for consideration of databases and comprehensive chemical characterization of 45 commercial juice samples from 23 different manufacturers in the United States reported by Zhang *et al.* (2009) [7].

Table 4. International Multidimensional Authenticity Algorithm Specifications (IMAS) for Commercial Pomegranate Juice [7].

Attribute (method)	IMAS criterion
Sugar profile	
glucose/fructose ratio (HPLC)	0.8-1.0
carbon SIRA ($^{13}C/^{12}C$ isotope ratio)	$\geq$ -25‰
mannitol (HPLC-RI)	$\geq$0.3 g/100 mL
sucrose (HPLC-RI)	not detectable
sorbitol (HPLC-RI)	$\geq$0.03 g/100 mL
maltose (HPLC-RI)	not detectable
Organic and amino acids	
citric/isocitric ratio (HPLC)	$\leq$350
tartaric acid (HPLC)	not detectable
malic acid (HPLC)	$\leq$0.1 g/100 mL
malic acid, D-isomer (HPLC)	not detectable
proline (HPLC)	$\leq$25 mg/L
Mineral	
potassium (flame photometer)	$\leq$1800 mg/L

Analytical Methodologies for Determination of Adulteration

The most frequent methods used to detect fruit juice adulteration are based on the profiling and quantification of some compounds that may be from one chemical family or different families, such as carbohydrates, phenolic compounds, amino acids, anthocyanins, pigments and organic acids.

Degrees of Brix

The simplest and the least expensive method of juice adulteration is dilution with water. This type of adulteration is simple to detect using a refractometer to determine the °Brix value of the juice [11]. As it is widely known, brix degree defines the percentage of water-soluble solids in fruit juice and can be affected by many factors such as variety, growth region, year and maturity level of the fruit. Relationships between brix and percentage sugars and acids can also be considered as parameters of composition of the juices and because of adulteration indices [12]. According to AIJN proposal, the minimum brix degree of pomegranate juice should be 14.0 [5].

Sorbitol/Xylitol Content

The difference on the sorbitol/xylitol content of fruit juices is an important parameter to detect juice adulteration. In order to confirm fruit juice adulteration, sorbitol/xylitol content is determined. The researches have shown that distribution of sugar alcohol is very distinctive among fruit species [5].

Characterization of Organic Acids Profile

Organic acids can be used as fingerprints in fruit juices, and are important to determine freshness and to detect adulteration. Ratios of organic acid often can be used to fingerprint a particular type of juice [13]. Tartaric acid is usually considered an indicator of grape juice addition to a more expensive juice (*e.g.,* pomegranate juice).

Analytical methods standard used for organic acids are based on liquid chromatography (reverse phase or ion exchange) coupled to UV detection. Various cooperative studies conducted by the Food Industry Analytical Chemists Committee of the Grocery Manufacturers Association have demonstrated that the inter-laboratory variability of results generated by these methods is often so high. However, accurate determination of the minor organic acids (*e.g.,* citric in apple, quinic acid in orange) has often been made a challenge [14].

Characterization of Amino Acids Profile

Especially, concentration and proportion of amino acids are sensitive indicators of authentication with relation to blending as well as fruit juices adulteration [9]. The addition of sugar syrup decreases the total amino acid value, and the use of constituents is not naturally present in the juice such as colorants and the addition of inexpensive juice from other types of a less expensive fruit. More sophisticated forms of adulteration consist in the use of cheap amino acids such as glycine or glutamic acid or protein hydrolysates to increase the total amino acid content. These adulterations can detect using the characterization of amino acids profile [15].

Characterization of Anthocyanin Profile and Polyphenol Composition

Chromatography combined with mass spectrometry has been used to successfully identify marker compounds of both pomegranate and the adulterant juices. These techniques require a certain amount of sample preparation before analysis and the chromatography process can be lengthy [9]. The HPLC method is highly sensitive and very fast in response. Separation efficiency is high. A wide range of compounds can be separated by HPLC, as the technique has a broad range of selectivity through the availability of many solvent combinations. The success of these methods depends on the presence of specific markers that might be present in either adulterant or pure juice [4].

Liquid chromatography (LC) is commonly used for anthocyanin characterization. Determination of free anthocyanins by liquid chromatography has been investigated to improve a method for quality control of these natural products found in red fruit juices, concentrated juices, and syrups. The anthocyanin pigment patterns exhibited by different fruit species have proved to be interesting for chemotaxonomic analysis and are now useful in order to control juice mixtures or juice adulteration [13 - 16]. UHPLC-PDA-fluorescence method was developed, capable of identifying and quantifying primary polyphenols able to detect and quantify the main polyphenols present in commercial fruit juices in a 28-min chromatogram [17].

The antioxidant capacity of commercial pomegranate juices is higher than that of several other polyphenolic-rich beverages such as tea and red wine. It has been attributed to the presence of hydrolysable tannins including gallotannins, the ellagitannins, punicalagin and punicalin. Pomegranates also contain anthocyanins with the main components being 3-O-glucosides and 3,5-O-diglucosides of cyanidin and delphinidin. The polyphenol composition of pomegranate is both complicated and unique. It can be used as a fingerprint by the food industry for quality control purposes. HPLC profiling with absorbance detection at 520 nm and fluorescence monitoring provide a simple and efficient way for accessing the authentication of pomegranate juices [18].

Stable Isotope Ratio Analysis (SIRA)

Stable isotope ratio analysis (SIRA) that has been applied for fruit juice is a method to determine authentication. This technique depends on the principle that naturally existing elements in various isotopic types and the distribution of these isotopes in the chemical compound profile of a fruit juice depends on a variety factors, such as growing region, temperature and photosynthetic pathway [11].

Fourier Transform Infrared (FTIR) Spectroscopy

The development of Fourier Transform Infrared (FTIR) spectroscopy working in the mid-infrared region has been created new perspectives in quality control for the food industry because it allows rapid screening and quantification of components and thus a high throughput of samples [19]. FTIR fingerprints technique has a significant advantage because of its rapidity and ease of spectral acquisition, enabling non-invasive measurements to be made with little or no sample preparation. FTIR-ATR provides a simple and reproducible means of handling products in the form of liquids and pastes, the analyses are non-destructive, and sampling takes less than 5 min. The FTIR method used in combination with chemometric techniques provides a promising approach for detecting adulteration of pomegranate juice concentrate with grape juice concentrate [4].

Oligosaccharide Profiling

Oligosaccharide profiling is another method for the detection of fruit juice adulteration. This technique works by identifying oligosaccharides that are present in the adulterants but are not naturally occurring in the product. One of the benefits of this method can be used to not only determine whether a product has been adulterated but also identify which adulterant was used. The most common techniques used for oligosaccharide profiling are high-performance anion-exchange chromatography with pulsed amperometric detection (HPAE-PAD) and capillary gas chromatography with flame ionization detection (GC-FID). Both HPAE-PAD and GC-FID methods have been developed and successfully used for detecting adulteration of carbohydrate-rich foods [11].

CONCLUSION

Fruit juice adulteration stands for an ongoing problem; suitable analytical methods are needed to control authentication. To date, some methods have been developed to address various aspects of fruit juice authentication. The most common approaches are based on profiling of carbohydrates, phenols, amino acids, or other organic acids using chromatographic and spectroscopic methods. Fingerprinting methods can provide rapid and high-throughput monitoring and would be ideally suited for rapid characterisations if considerable changes could capture in a reproducible way. These protocols ensure the required information and will be useful for monitoring pomegranate juice adulteration.

CONSENT FOR PUBLICATION

Not applicable.

CONFLICT OF INTEREST

The author (editor) declares no conflict of interest, financial or otherwise.

ACKNOWLEDGEMENT

Declared none.

REFERENCES

[1]   Navarro-Pascual-Ahuir M, Lerma-García MJ, Simó-Alfonso EF, Herrero-Martínez JM. Quality control of fruit juices by using organic acids determined by capillary zone electrophoresis with poly(vinyl alcohol)-coated bubble cell capillaries. Food Chem 2015; 188: 596-603.
[http://dx.doi.org/10.1016/j.foodchem.2015.05.057] [PMID: 26041236]

[2]   Les F, Prieto JM, Arbonés-Mainar JM, Valero MS, López V. Bioactive properties of commercialised pomegranate (Punica granatum) juice: antioxidant, antiproliferative and enzyme inhibiting activities. Food Funct 2015; 6(6): 2049-57.
[http://dx.doi.org/10.1039/C5FO00426H] [PMID: 26030005]

[3]   Tastan O, Baysal T. Clarification of pomegranate juice with chitosan: changes on quality characteristics during storage. Food Chem 2015; 180: 211-8.
[http://dx.doi.org/10.1016/j.foodchem.2015.02.053] [PMID: 25766820]

[4]   Vardin H, Tay A, Ozen B, Mauer L. Authentication of pomegranate juice concentrate using FTIR spectroscopy and chemometrics. Food Chem 2008; 108(2): 742-8.
[http://dx.doi.org/10.1016/j.foodchem.2007.11.027] [PMID: 26059156]

[5]   Türkmen I, Ekşi A. Brix degree and sorbitol/xylitol level of authentic pomegranate (Punica granatum) juice. Food Chem 2011; 127(3): 1404-7.
[http://dx.doi.org/10.1016/j.foodchem.2010.12.118] [PMID: 25214145]

[6]   Nuncio-Jáuregui N, Calín-Sánchez Á, Hernández F, Carbonell-Barrachina AA. Pomegranate juice adulteration by addition of grape or peach juices. J Sci Food Agric 2014; 94(4): 646-55.
[http://dx.doi.org/10.1002/jsfa.6300] [PMID: 23847043]

[7]   Zhang Y, Krueger D, Durst R, *et al.* International multidimensional authenticity specification (IMAS) algorithm for detection of commercial pomegranate juice adulteration. J Agric Food Chem 2009; 57(6): 2550-7.
[http://dx.doi.org/10.1021/jf803172e] [PMID: 19249817]

[8]   Anon.AIJN-provisional reference guideline for pomegranate juice. European Fruit Juice Association, 1990, Brussels.

[9]   Singhal RS, Kulkarni PR, Rege DV. Handbook of indices of food quality and authenticity. Woodhead Publishing Ltd 1997; p. 561.
        [http://dx.doi.org/10.1533/9781855736474]

[10]   Ekşi A, Özhamamcı İ. `Chemical composition and guide values of pomegranate juice. Gida 2009; 34(5): 265-70.

[11]   Willems JL, Low NH. Authenticity analysis of pear juice employing chromatographic fingerprinting. J Agric Food Chem 2014; 62(48): 11737-47.
        [http://dx.doi.org/10.1021/jf5043618] [PMID: 25384245]

[12]   Maireva S, Manhokwe S. The Determination of Adulteration in Orange Based Fruit Juices. International Journal of Science and Technology 2013; 2(5): 365-72.

[13]   LaCourse WR. Ion Chromatography in Food Analysis, Handbook of Food Analysis Instruments CRC Press. 2009; p. 188.

[14]   Ehling S, Cole S. Analysis of organic acids in fruit juices by liquid chromatography-mass spectrometry: an enhanced tool for authenticity testing. J Agric Food Chem 2011; 59(6): 2229-34.
        [http://dx.doi.org/10.1021/jf104527e] [PMID: 21361392]

[15]   Asadpoor M, Ansarin M, Nemati M. Amino Acid profile as a feasible tool for determination of the authenticity of fruit juices. Adv Pharm Bull 2014; 4(4): 359-62.
        [PMID: 25436191]

[16]   Goiffon JP, Mouly PP, Gaydou EM. Anthocyanic pigment determination in red fruit juices, concentrated juices and syrups using liquid chromatography. Anal Chim Acta 1999; 382: 39-50.
        [http://dx.doi.org/10.1016/S0003-2670(98)00756-9]

[17]   Díaz-García MC, Obón JM, Castellar MR, Collado J, Alacid M. Quantification by UHPLC of total individual polyphenols in fruit juices. Food Chem 2013; 138(2-3): 938-49.
        [http://dx.doi.org/10.1016/j.foodchem.2012.11.061] [PMID: 23411199]

[18]   Borges G, Crozier A. HPLC-PDA-MS fingerprinting to assess the authenticity of pomegranate beverages. Food Chem 2012; 135(3): 1863-7.
        [http://dx.doi.org/10.1016/j.foodchem.2012.05.108] [PMID: 22953935]

[19]   Fügel R, Carle R, Schieber A. Quality and authenticity control of fruit purees, fruit preparations, and jams-a review. Trends Food Sci Technol 2005; 16: 433-41.
        [http://dx.doi.org/10.1016/j.tifs.2005.07.001]

| CHAPTER 7 |

Adulteration of Synthetic Substances in Dietary Supplements

Nicholas Schramek * and **Uwe Wollein**

Bavarian Health and Food Safety Authority, Oberschleißheim, Germany

Abstract: The use of dietary supplements is widespread. Concurrent adulteration of such products with undeclared pharmaceuticals has been reported quite often, especially for 'the big three', erectile dysfunction, weight-loss, and sports performance products. But still most consumers are unfamiliar with the problem, posing potentially serious health consequences. This article highlights the most common synthetic substances found as adulterants in dietary supplements, describes the analysis and the difficulties with the scientific assessment, using sexual enhancement products as an example.

Keywords: Adulteration, Anabolic steroids, Dietary supplement, Doping, Erectile Dysfunction drugs, Food laws, PDE5 inhibitors, Slimming products, Sildenafil, Sibutramine, Sexual enhancement products.

INTRODUCTION

Dietary supplements are commonly used among the whole population [1]. They are commonly regulated by food law and therefore can be placed on the market – which is often worldwide, as many of these products are sold *via* internet– without control by competent authorities. Although they are supposed to comply with the relevant requirements of food law it turns out, that many of these products do not. Adulteration of dietary supplements with active pharmaceutical ingredients or designer drugs are significant and has been on the rise for years [2, 3]. This is especially true for the popularly called 'big three', erectile dysfunction, weight loss, and bodybuilding/sport performance products [4]. Not infrequently, these product groups are advertised as 'pure natural', containing only a blend of fine selected herbs, but in reality they are adulterated with one or more non declared active ingredients.

---

* **Corresponding author Nicholas Schramek:** Bavarian Health and Food Safety Authority, Oberschleißheim, Germany; Tel: +49 9133 6808 5451; E-mail: nicholas.schramek@lgl.bayern.de

Alankar Shrivastava (Ed.)

As early as 1997 Huang *et al.* found that about 24% of 2609 samples of traditional Chinese medicines analyzed in Taiwan were adulterated with synthetic drugs of various pharmacological activities [5]. Analysis of weight loss products in the Netherlands from 2002 – 2007 showed that most of them were adulterated [6]. Many people do not (or in some cases don´t really want to) recognize the risks associated with dietary supplements, which pose potentially serious public health consequences [7].

SEXUAL ENHANCEMENT PRODUCTS

Sexual enhancement products are widely sold *via* internet shops, but are also available in erotic stores. They are advertised as pure natural, containing selected herbs like, *e.g.* Epimedium, Schisandra, Ginkgo, Tribulus, Maca, or Lycium fruit. These products were sold outside the official health system as dietary supplements, whereas evaluations imply these sales would be 2.5 times greater than legitimate prescriptions of authorized medicines for treatment of erectile dysfunction (ED) like Viagra® [8]. Different market surveillance studies established over the national and even the EU-wide market show, that 38% (national) or 50% (EU-wide) of the analyzed samples contain undeclared APIs [authors personal data]. Most of the adulterations found consist of sildenafil or one of its structural analogues (Fig. **1**), more seldom tadalafil (API of Cialis®), and very rarely vardenafil (API of Levitra®).

Nevertheless, structural related analogues from the latter compounds were found. Known tadalafil analogues differ only at the nitrogen from the pyrazine-ring, including one *N*-amino derivative, and one ring-open precursor (chloropretadalafil), which is leading to further *N*-alkyl compounds, depending on the use of *R*-amino reagents (Fig. **2**) [9 - 12].

Fewer structures of vardenafil analogues (Fig. **2**) [13 - 15] could be found in literature, which might be explained by the pharmacological profile of vardenafil against sildenafil: action occurs faster by a shorter duration.

The first reported analogue of sildenafil was homosildenafil in 2003 [16], followed by acetildenafil (hongdenafil) in 2004 [17]. Within 12 years of analytical investigations, more than 70 analogue structures have been found and published in scientific literature [18], and the number is still growing (for reviews see [19 - 21]). Analogues are by-products of the original drug discovery process carried out by pharmaceutical industry.

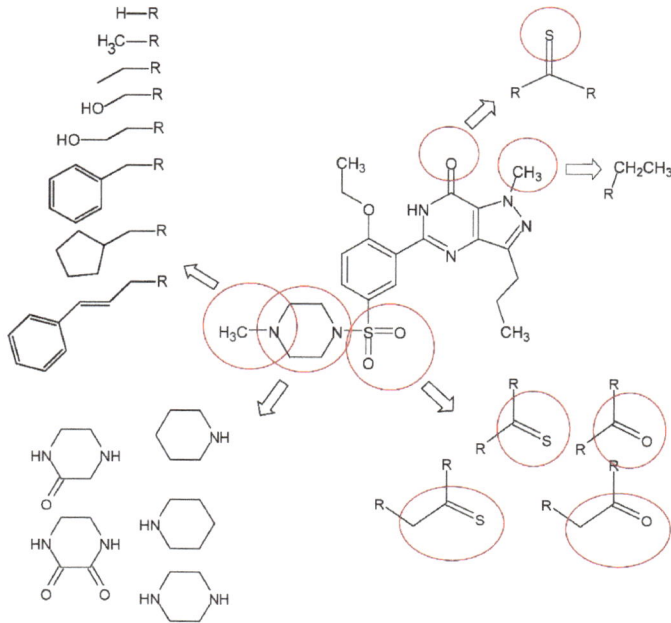

Fig. (1). Sildenafil and some of the possible modifications in the chemical structure found in the literature.

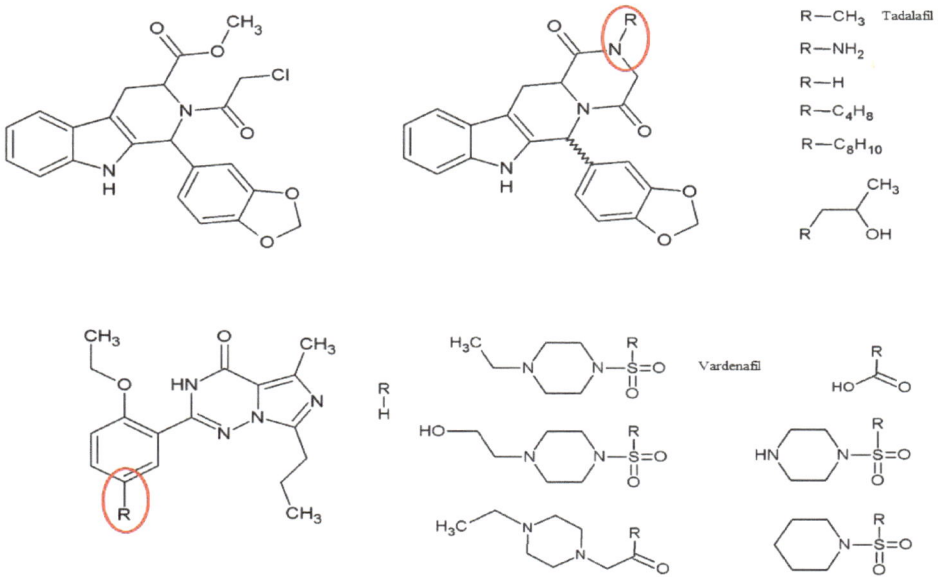

Fig. (2). Literature known analogues of tadalafil and vardenafil.

Detailed synthesis steps towards these structures and preliminary pharmacology can nowadays be easily retraced from the patent literature [22 - 26]. Fig. (**1**) presents the main sites of the sildenafil basic structure (5-(2-Ethoxyphenyl)-1 - methyl-3-propyl-1,6-dihydro-*7H*-pyrazolo[4,3-d]pyrimidine-7-one)   to   be modified by chemical synthesis. Thio-analogues show a pyrazolo[4,3-d]pyrimidine-7-thione basic structure and can be synthesized, *e.g.* by heating sildenafil with P_2S_5. Thio-derivatives are supposed to have a lower IC_{50}-value in PDE5 inhibition as their oxo precursors [27], but are not registered as drug substances at the EMA or FDA. Another structure modification is leading to acetildenafil-type analogues, where the SO_2 group is replaced by a $-CH_2CO-$ bridge. Acetildenafil is supposed to show only 90% of sildenafil's activity with regard to their IC_{50}-values [28]. More exotic structures linking the phenyl ring with the piperazine moiety *via* a carbonyl or thiocarbonyl group only, generate compounds like carbodenafil [29] or dithiodesmethylcarbodenafil [30]. Even dihydro derivatives of acetildenfil were detected, but showing a carbonyl function in the piperazine ring system [31]. The easiest modifications could be established within the piperazine ring, with primal focus on the 4'-nitrogen. Findings reach from unsubstituted nitrogen over alkyl to aryl groups. A commonly found analogue bears a hydroxyethyl-moiety leading to the hydroxyhomo-prefix in the trivial name of the sildenafil related compound [32, 33]. Also dimethylpiperazine is detected and named as aildenafil, or thioaildenafil as the thioketone analogue [34].

Sexual enhancement products aim not only at men but also at women. PDE5 inhibitors do not have an effect on stimulating women's sexual desire, but flibanserin– only recently approved by the FDA – was found to influence pre-menopausal women with hypoactive sexual desire disorder. It has also been detected as a synthetic adulterant in herbal food supplements [35, 36].

Methods of Analysis

Methods used for the detection and identification of ED-drugs and their analogues cover almost the whole range of analytical techniques. To perform a low cost pre-screening, Thin Layer Chromatography (TLC) should not be disregarded beside its modern and hyphenated alternatives. Using commonly available GF 254 silica gel plates, *e.g.* a mobile phase consisting of chloroform/methanol/diethylamine (90:10:1 [v/v/v]), presumably all of the relevant compounds can be detected under UV light. Furthermore, when using a suitable reagent, some preliminary information for identification of adulterations may be achieved. Thio-derivatives of sildenafil appear as a bright yellow spot when spraying the plate with iodoplatinate reagent, whereas the oxo-compounds yield a brown color. Acetildenafil and its analogues are developing a more brownish or black color

(Fig. **3**). On the other hand, when using sulphuric acid as a visualization reagent, tadalafil appears markedly purple color after heating the plate at 100 °C. Similar rapid screening results with TLC are also described in literature [37 - 40].

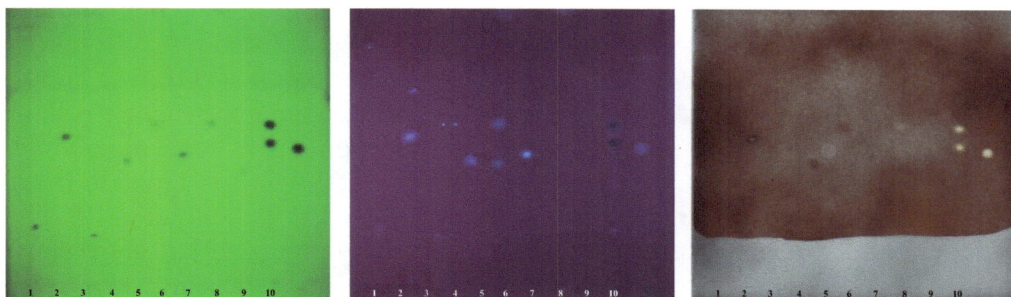

Fig. (3). Thin layer chromatography of selected adulterations at 254 nm, 365 nm and with iodoplatinate reagent (1: Acetildenafil, 2: Homosildenafil, 3: Hydroxyacetildenafil, 4: Sildenafil, 5: Tadalafil, 6: Vardenafil, 7: Dithiodesmethylcarbodenafil, 8: Nitrosoprodenafil (Mutaprodenafil), 9: Hydroxyhomothiosildenafil + Thiosildenafil, 10: Flibanserin).

High performance liquid chromatography (HPLC) coupled with a diode array detector (HPLC-DAD) is an easy and relatively low-cost technique for the detection of illegal compounds in food. Not only the retention time is available for characterization of a compound, but also the UV-spectrum, typically recorded between 200 and 400 nm. Analogues may be differentiated by their absorbance maxima, as the sildenafil-group (221 nm, 295 nm) is significant deviating from its thio-analogue-group (225 nm, 295 nm, and 354 nm), or the acetildenafil-group (232 nm, 282 nm) [29]. With the use of reference standards, a library can be created in order to identify the adulterant [41].

LC-MS based methods receive more and more attention in the recent years. Modern ionization techniques, *e.g.* electro spray ionization (ESI) in positive or negative mode are frequently used. Tandem or multi-stage mass spectrometry (MS/MS or MS^n) results in either higher resolution on the origin of the observed fragment ions of the possibly unknown structure, or higher sensitivity when screening for unknown structures at a low level [42]. Sildenafil analogues are marked by fragment ions at *m/z* 377, 311, and 283. Thiosildenafil shows a characteristic daughter ion at *m/z* 299, whereas acetildenafil typically fragments into *m/z* 353, 396, and 449. Interestingly, when using ESI in negative mode, a further structural information is given by the neutral loss of ethene from the ethoxy moiety of the basic structure, represented by a [(M-1)-28]$^-$ from the corresponding parent ion [43].

If all the above mentioned data are still insufficient for structure elucidation, nuclear magnetic resonance (NMR) analysis is the method of choice, especially for novel compounds. With today's available guidance, characterization of analogues by characteristic signals becomes very easy [44]. For the sildenafil basic structure, the aromatic region between 8.5 and 7 ppm of the ^{1}H-NMR spectra and 200 and 110 ppm of the ^{13}C-NMR spectra, respectively, appears very similar. These high field regions represent the 1,2,4-substituted phenyl ring (^{1}H) in combination with the pyrazolopyrimidinone substructure (^{13}C). The methyl group bonded to the nitrogen of the pyrazolopyrimidinone substructure reliably appears as a singlet at about 4.5 ppm (^{1}H) and 40 ppm (^{13}C), respectively. The signals at about 4.3 ppm (quartet) and 1.7 ppm (triplet) can be assigned to the O-ethyl moiety, whereas the signals at about 2.9 (double doublet), 1.8 ppm (multiplet), and 1.0 ppm (triplet) represent the 1-propyl group [43]. Fig. (**4**) shows typical ^{1}H-NMR signals from three different analogues [18] in comparison with the basic pyrazolopyrimidinone structure of sildenafil [43].

For further structure elucidation, two-dimensional NMR techniques have to be employed. Homonuclear (*e.g.* COSY, NOESY) and heteronuclear (HSQC, HMBC) are the most frequently used experiments.

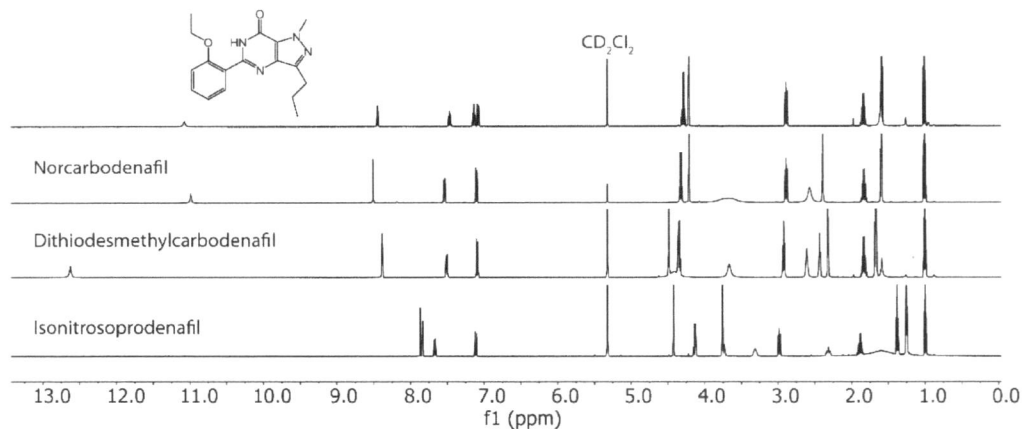

Fig. (4). ^{1}H-NMR spectra of norcarbodenafil, dithiodesmethylcarbodenafil, and isonitrosoprodenafil, compared with the ^{1}H-NMR spectrum of the basic pyrazolopyrimidinone structure of sildenafil.

Further analytical methods to characterize PDE5 inhibitors and their analogues that are discussed in the literature include are near infrared (NIR) coupled with chemometric tools [45], ion mobility spectrometry (IMS) [46], gas chromatography coupled with mass spectrometry (GC-MS) [47], and immuno assay methods [48, 49].

Quantification of PDE5 inhibitors and their analogues is commonly carried out by (U)HPLC methods and UV or mass detection using reference substances. Many of the PDE5 analogues are commercially available, but most of them are, however, very expensive. As a compromise, inexpensive sildenafil is frequently used as the only external standard together with a HPLC-UV method. Correction factors may be used when known and found to be necessary [50]. Although this is leading to an inexact quantification of the target substance, it should nevertheless be sufficient for legal evaluation against applicable law. If no reference standard is available and an NMR is at hand, the quantification of nearly all compounds is possible using qNMR techniques [51].

Assessment of Sexual Enhancement Products

When sexual enhancement products are sold as food supplements, analysts should be aware that these products often lack commonly accepted GMP. This leads to difficulties that may arise during assessment. Within a single product, individual capsules can differ significantly from each other with regard to their composition (Fig. **5**) [43].

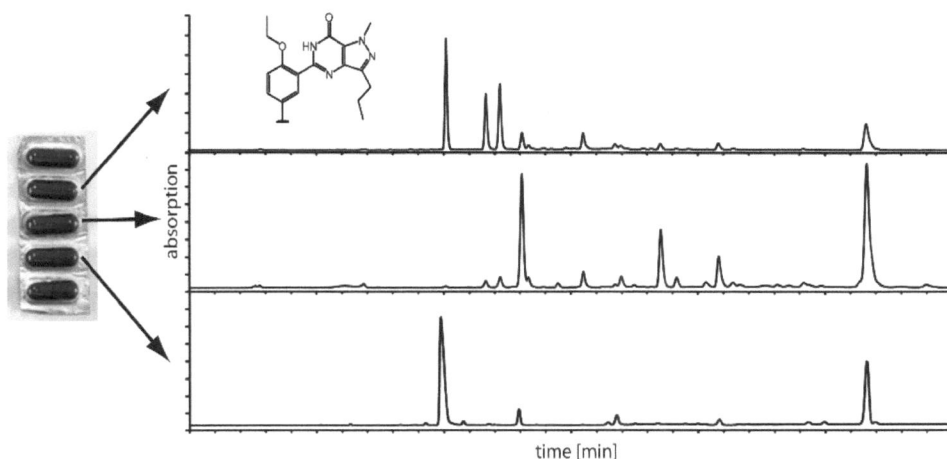

Fig. (5). Significant differences between single capsules out of the same product. All peaks could be assigned to compounds bearing the sildenafil basic structure.

Moreover, when sampling the same product at, *e.g.* different times or from different sources, the analytical results may be completely different, even by identical labelling (*e. g.* same Lot number and expiry date). It could be found, that packages identical in appearance and declaration contain very different adulterants, while others contain no illegal compounds at all (Table **1**).

Table 1. Examples of products with the same labelling but with different contents (numbers bracketed show either number of dosage forms per package or detected amount in mg per form); data obtained either from own investigations or KnowX database of the EDQM.

Product	Form	Adulteration
Man Power	Bright blue capsules (10)	Dithiodesmethylcarbodenafil (57 – 67)
Man Power	Bright blue capsules (2)	Nitrosoprodenafil/Mutaprodenafil/Nitroprodenafil (188)
Rock Hard weekend	n/a	Thioaildenafil (82)
Rock Hard weekend	Blue and white capsule (1)	Aildenafil (n/a)
Rock Hard weekend	Blue and white capsule (1)	No sildenafil derivate found (contains caffeine)
Power tabs	Bright blue capsules (10)	Dithiodesmethylcarbodenafil (50)
Power tabs	n/a	Dithiodesmethylcarbodenafil and Thioaildenafil
Power tabs	Bright blue capsules (10)	N-Desmethylacetildenafil and Dithiodesmethylcarbodenafil (n/a)
Power tabs	Bright blue capsules (10)	Dithiodesmethylcarbodenafil (100) and minor components from different synthetic routes (Isonitrosoprodenafil, Dithiodesethylcarbodenafil, Norcarbodenafil)
Power tabs	Bright blue capsules (20)	N-Butyltadalafil (40)
Viamax pure power	Bright blue capsules (10)	No sildenafil derivate found
Viamax pure power	Bright blue capsules (10)	Dithiodesmethylcarbodenafil (25)
Viamax desire	Red capsules (2)	Flibanserin (4)
Viamax desire	Red capsules (2)	No adulteration found

For some of the PDE5 analogues in-vitro data may be available, that may further lead to a toxicological assessment, whilst others are completely unknown. Some may be synthesis by-products; some may bear reactive functional groups which make assessors assuming a higher potential risk for patients. Sometimes the adulterants are derived from different synthetic pathways, or they may be swept from the benchtop of clandestine laboratories directly into capsules without any further purification. With all the above said, an assessment of these suspect products is not possible at any time and will lead to false negative screening in the worst case. This may lead accordingly to inaccurate risk assessment, as the product may cause harm to other patients [52]. Significant side effects after use of these food supplements are well reported. They were described as headaches, facial flushing, nasal congestion, and visual distortions. Further effects sourcing from a thio-analogue of sildenafil in a food supplement were vertigo, headache, shortness of breath and backache [53].

SLIMMING PRODUCTS

Adulterated products for weight loss comprise of large group of very different substances, including stimulants, antidepressants, laxatives, and diuretics [4, 54 - 60] (Table **2**). Most of them are banned or withdrawn from the market due to health risks or are only available on prescription. Ephedrin/*Ephedra*-containing dietary supplements for example were used years for gaining weight loss. Through 2001 there were more than 13,000 health complaints and 100 deaths assigned to ephedrine [61]. In 2004 an ever-increasing number of adverse effects prompted the FDA to remove these products from the market [62].Taking slimming products is regarded by most people as being harmless due to their natural origin or safe composition (according to the product label) [63]. But indeed their intake can be far riskier compared to prescription drugs and they are known to be linked with many different adverse health effects including death [64, 65].

Table 2. Examples of compounds found as adulterations in weight loss products.

Substance Type	Compound examples
Stimulant	- ephedrine - norephedrine - pseudoephedrine - fenfluramine - phentermine - sibutramine -*N*-desmethylsibutramine -*N,N*-didesmethylsibutramine - rimonabant
Antidepressant	- doxepin - fluoxetine - sertraline
Diuretic	- hydrochlorothiazide - methylclothiazide - chlorothiazide - furosemide
Laxative	- bisacodyl - phenolphthalein

(Table 2) contd.....

Substance Type	Compound examples
Lipase inhibitor	- orlistat
Other	- 2,4-dinitrophenol

Sibutramine (Fig. **6**) may be one of the most commonly found substance not only in adulterated supplements but also in "slimming coffee" and similar products for weight loss [66 - 69]. It acts as a serotonine-noradrenaline reuptake inhibitor and stimulates thermogenesis [70]. After approval for weight loss and maintenance of weight loss in obese people, as well as in certain overweight people with other risks for heart disease in 1997 in the US, sibutramine was withdrawn from the market in 2010 in several countries due to the risk of serious cardiovascular events [71, 72].

Besides sibutramine itself, various analogues like *N*-desmethylsibutramine, *N,N*-didesmethylsibutramine, benzylsibutramine, and chlorosibutramine have been found as adulterants insupplements (Fig. **6**) [52, 73 - 76].

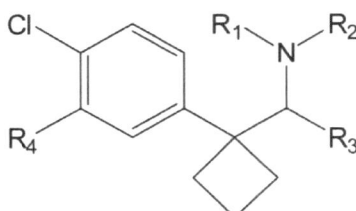

R₁	R₂	R₃	R₄	Compound
Me	Me	Isobutyl	H	Sibutramine
H	Me	Isobutyl	H	Desmethyl-sibutramine
H	H	Isobutyl	H	Didesmethyl-sibutramine
Formyl	H	Isobutyl	H	Formylsibutramine
Me	Me	Isobutyl	Cl	Chlorosibutramine
Me	Me	Benzyl	H	Benzylsibutramine

Fig. (6). Chemical structure of sibutramine and some of its analogues.

Rimonabant, a cannabinoid receptor 1 (CB1) antagonist, is another anti-obesity drug found in weightloss products [56, 77 - 79]. It was sold in parts of the European Union as a prescription drug since 2007, but as early as 2008 it was suspended due to the risks of dangerous psychological side effects (including suicidality) and was finally withdrawn from the market in the beginning of 2009

[80]. Rimonabant has never been approved by the FDA. Within the EU market it was available between 2006 and 2008 under the trade name Acomplia®.

2,4-Dinitrophenol (DNP) is used as a weight loss agent targeted towards primarily body builders. Originally used in the manufacturing of dyes, wood preservatives, insecticides, and explosives, it found widespread demand in the early 1930s as a weight loss agent [81]. DNP acts by uncoupling oxidative phosphorylation in mitochondria. Studies showed that a daily dose of 300 – 400 mg of DNP can nearly double the basal metabolic rate of healthy subjects [82]. In 1938 DNP was taken off the market as a result of adverse effects including cataracts, liver failure, agranulocytosis, skin toxicity, and death [81, 83 - 89]. It reappeared in the 1980s as a slimming agent [90]. The lethal dose is estimated to be 1 – 3 g taken orally. Moreover, 3 g of DNP has been shown to be fatal, even if the dose is divided and taken over a period of five days [91]. Cases of deaths were reported even following administration within the "recommended" dosage [87], offered on a wide range of websites.

Phenolphthalein, usually known as a visual indicator in titrations, has also been used as an over-the-counter laxative worldwide [92]. It was removed from the FDA list of products generally recognized as safe and effective for use as laxative drug in 1999 [93]. Phenolphthalein is now considered to be a human carcinogen. Nevertheless it is still found in dietary supplements [57 - 59, 78, 79, 94].

Other common adulterants in weight loss products are diuretics (*e.g.* hydrochlorothiazide, furosemide) [65, 95], appetite suppressants (*e.g.* phentermine, fenfluramine) [64, 65, 96], antidepressants (e. g. sertraline, fluoxetine) [97], or the lipase inhibitor orlistat [56, 98, 99].

BODYBUILDING AND SPORT PERFORMANCE PRODUCTS

Bodybuilding and sport performance products comprise very different forms, *i. e.* energy bars, isotonic drinks, tablets, injectable solutions, or (herbal) dietary supplements. They are taken by sportsperson and athletes of all levels to improve performance or to gain muscle and increase muscle growth [100]. In recent years it has been shown, that dietary supplements contain unlabeled substances, most of them banned by the World Anti-Doping Agency (WADA). Nearly all kinds of active agents are conceivable as adulterants but anabolic androgenic steroids are the most common ones [101 - 104]. Even if they are present in very low amounts as shown in an international study by Geyer *et al.* in 2004, this may lead to positive results in doping tests [103]. Also supplements marketed as 'legal prohormones', often contain well established anabolic steroids [103, 105]. Mislabeling is commonly found within these product group (Table **3**). But even if they contain the labeled 'prohormones', little – if anything – is known about the toxicity or efficacy of these substances [104].

Table 3. Examples of 'prohormone supplements', where labelling did not reflect their actual content [authors personal data].

Product	Declared ingredient(s)	Actual content
Finaflex 1-Andro	• 3-Hydroxy-5a-androst-1-en-17-one • Bergamottin (6,7-dihydroxy-bergamottin) (DHB)	• 1-Androsterone (about 33 mg/capsule) • DHEA (about 29 mg/capsule)
EPI Brawn Nutrition	2a,3a-Epithio-17a-methyl-5a-androstan-17b-ol (10 mg/capsule)	• Metasterone (about 2 mg/capsule) • Dymethazine (about 10 mg/capsule)
Halo Increase Lean Muscle	Chloro-17a-methyl-4-ene-3,17-diol (25 mg/capsule)	4-Chlorodehydromethyltestosterone –‚Oral-Turinabol' (27,5 mg/capsule)

OTHER LIFESTYLE PRODUCTS

Adulterants can be found in a wide range of other lifestyle or 'health related' products. Non-steroidal inflammatory drugs like phenylbutazone or diclofenac [75, 106] as well as steroid compounds [107] have been found in herbal remedies against rheumatism and pain. Glibenclamide and other antidiabetic drugs are common adulterants in dietary supplements against diabetes [106]. Illegal adulteration with sedative hypnotics such as diazepam, nitrazepam and other benzodiazepines have been routinely detected in dietary supplements and Traditional Chinese Medicines [108]. Dietary supplements for blood pressure or blood lipid management have been found to contain β-blockers, statins, ACE inhibitors and other pharmaceuticals [109, 110]. Undeclared prostaglandine derivates (*e.g.* latanoprost, bimatoprost, isopropyl cloprostenate) have been identified in products for eyelash growth [99, 111]. Other cosmetics, like creams and ointments for skin conditions declared as active-agent free, frequently contain corticosteroids like betamethasone or dexamethasone [99].

CONCLUDING REMARKS

In most people's minds, dietary supplements, especially the ones declared 'all-natural', are regarded to be safe and to bring only positive effects. However, adulteration of dietary supplements and other foods or even herbal medicines with synthetic drugs for obtaining faster and better effects is a very common problem [112]. Since the 'real' active ingredient(s) and its/their dose are/is unknown to the user, there can be a serious health risk. But even if a product has been analyzed and found to be free from adulterations, it has been shown that there can be fundamental differences in the composition of dosage units within one package of an adulterated supplement and also between samples taken from different sources

[43, 52]. Probably many dietary supplements on the market sold *e.g.* for vitamin or mineral supplementation, are both safe and of high-quality. On the other hand, there are numerous diet and other lifestyle supplements of questionable quality, safety, and health benefit. For consumers it is hard to get definite knowledge about the quality. If there is a tangible effect after administration of a certain product, it is likely due to an illegally added substance whose safety is unproven – or in other words: 'If your supplement works amazingly fast … it´s bad for you' [113]. As long as there is no adequate regulation of these 'medicine-like' food supplements, very high risk for the consumers remains, that is unacceptably high especially for certain at-risk groups.

CONSENT FOR PUBLICATION

Not applicable.

CONFLICT OF INTEREST

The author (editor) declares no conflict of interest, financial or otherwise.

ACKNOWLEDGNEMT

Declared none.

REFERENCES

[1]   Murphy SP, Wilkens LR, Monroe KR, *et al.* Dietary supplement use within a multiethnic population as measured by a unique inventory method. J Am Diet Assoc 2011; 111(7): 1065-72.
        [http://dx.doi.org/10.1016/j.jada.2011.04.004] [PMID: 21703385]

[2]   Deconinck E, De Leersnijder C, Custers D, Courselle P, De Beer JO. A strategy for the identification of plants in illegal pharmaceutical preparations and food supplements using chromatographic fingerprints. Anal Bioanal Chem 2013; 405(7): 2341-52.
        [http://dx.doi.org/10.1007/s00216-012-6649-4] [PMID: 23307125]

[3]   Cole MR, Fetrow CW. Adulteration of dietary supplements. Am J Health Syst Pharm 2003; 60(15): 1576-80.
        [PMID: 12951758]

[4]   Sarma   N.   http://www.nutritionaloutlook.com/articles/dietary-supplement-adulteration-erectile-dysfunction-weight-loss-and-sports2015.

[5]   Huang WF, Wen KC, Hsiao ML. Adulteration by synthetic therapeutic substances of traditional Chinese medicines in Taiwan. J Clin Pharmacol 1997; 37(4): 344-50.
        [http://dx.doi.org/10.1002/j.1552-4604.1997.tb04312.x] [PMID: 9115061]

[6]   Venhuis BJ, Zwaagstra ME, van den Berg JDJ, *et al.* Trends in drug substances detected in illegal weight-loss medicines and dietary supplements; RIVM Report 370030002/2009; National Institute for Public Health and the Environment: 2009.

[7]   Cohen PA. American roulette--contaminated dietary supplements. N Engl J Med 2009; 361(16): 1523-5.
        [http://dx.doi.org/10.1056/NEJMp0904768] [PMID: 19812394]

[8]   Sugita M, Miyakawa M. Economic analysis of use of counterfeit drugs: health impairment risk of counterfeit phosphodiesterase type 5 inhibitor taken as an example. Environ Health Prev Med 2010; 15(4): 244-51.
[http://dx.doi.org/10.1007/s12199-010-0134-5] [PMID: 21432552]

[9]   Zou P, Hou P, Oh SS-Y, Low M-Y, Koh H-L. Electrospray tandem mass spectrometric investigations of tadalafil and its analogue. Rapid Commun Mass Spectrom 2006; 20(22): 3488-90.
[http://dx.doi.org/10.1002/rcm.2752] [PMID: 17066376]

[10]   Hasegawa T, Takahashi K, Saijo M, *et al.* Isolation and structural elucidation of cyclopentynafil and N-octylnortadalafil found in a dietary supplement. Chem Pharm Bull (Tokyo) 2009; 57(2): 185-9.
[http://dx.doi.org/10.1248/cpb.57.185] [PMID: 19182409]

[11]   Toomey VM, Litzau JJ, Flurer CL. Isolation and structural characterization of two tadalafil analogs found in dietary supplements. J Pharm Biomed Anal 2012; 59(0): 50-7.
[http://dx.doi.org/10.1016/j.jpba.2011.09.038] [PMID: 22055930]

[12]   Lee JH, Kim HJ, Noh E, *et al.* Identification and screening of a tadalafil analogue found in adulterated herbal products. J Pharm Biomed Anal 2015; 103(0): 80-4.
[http://dx.doi.org/10.1016/j.jpba.2014.11.006] [PMID: 25462124]

[13]   Park HJ, Jeong HK, Chang MI, *et al.* Structure determination of new analogues of vardenafil and sildenafil in dietary supplements. Food Addit Contam 2007; 24(2): 122-9.
[http://dx.doi.org/10.1080/02652030600983625] [PMID: 17364912]

[14]   Lam Y-H, Poon W-T, Lai C-K, Chan AY-W, Mak TW-L. Identification of a novel vardenafil analogue in herbal product. J Pharm Biomed Anal 2008; 46(4): 804-7.
[http://dx.doi.org/10.1016/j.jpba.2007.12.004] [PMID: 18248930]

[15]   Reepmeyer JC, Woodruff JT. Use of liquid chromatography-mass spectrometry and a hydrolytic technique for the detection and structure elucidation of a novel synthetic vardenafil designer drug added illegally to a "natural" herbal dietary supplement. J Chromatogr A 2006; 1125(1): 67-75.
[http://dx.doi.org/10.1016/j.chroma.2006.05.018] [PMID: 16750214]

[16]   Shin MH, Hong MK, Kim WS, Lee YJ, Jeoung YC. Identification of a new analogue of sildenafil added illegally to a functional food marketed for penile erectile dysfunction. Food Addit Contam 2003; 20(9): 793-6.
[http://dx.doi.org/10.1080/0265203031000121455] [PMID: 13129773]

[17]   Blok-Tip L, Zomer B, Bakker F, *et al.* Structure elucidation of sildenafil analogues in herbal products. Food Addit Contam 2004; 21(8): 737-48.
[http://dx.doi.org/10.1080/02652030412331272467] [PMID: 15370823]

[18]   Schramek N, Wollein U, Eisenreich W. Identification of new synthetic PDE-5 inhibitors analogues found as minor components in a dietary supplement. J Pharm Biomed Anal 2014; 96: 45-53.
[http://dx.doi.org/10.1016/j.jpba.2014.03.023] [PMID: 24726888]

[19]   Venhuis BJ, de Kaste D. Towards a decade of detecting new analogues of sildenafil, tadalafil and vardenafil in food supplements: a history, analytical aspects and health risks. J Pharm Biomed Anal 2012; 69(0): 196-208.
[http://dx.doi.org/10.1016/j.jpba.2012.02.014] [PMID: 22464558]

[20]   Patel DN, Li L, Kee C-L, Ge X, Low M-Y, Koh H-L. Screening of synthetic PDE-5 inhibitors and their analogues as adulterants: analytical techniques and challenges. J Pharm Biomed Anal 2014; 87(0): 176-90.
[http://dx.doi.org/10.1016/j.jpba.2013.04.037] [PMID: 23721687]

[21]   Singh S, Prasad B, Savaliya AA, Shah RP, Gohil VM, Kaur A. Strategies for characterizing sildenafil, vardenafil, tadalafil and their analogues in herbal dietary supplements, and detecting counterfeit products containing these drugs. Trends Analyt Chem 2009; 28(1): 13-28.
[http://dx.doi.org/10.1016/j.trac.2008.09.004]

[22]   Bell A S, Brown D, Terrett N K. Preparation of pyrazolo[4,3-d]pyrimidin-7-ones as cardiovascular agents. EP463756A1, 1992.

[23]   Ellis P, Terrett N K. Pyrazolopyrimidines for the treatment of impotence. WO94/28902, 1994.

[24]   Campbell S, Mackenzie A, Wood A. cGMP-PDE inhibitors for the treatment of erectile dysfunction. US20040087599 A1, 2004.

[25]   Kim J H, Kim Y, Choi D H, Kim D H, Nam G H, Seo J H. Novel pyrazolopyrimidinethione derivatives, preparation methods thereof and their use as therapeutics for erectile dysfunction. US0176371A1, 2004.

[26]   Piazza G A, Pamukcu R. Preparation of phenyl purinone derivatives for the treatment of precancerous lesions. US6200980B1, 2001.

[27]   Cho E-Y, Chung S-H, Kim J-H, Kim D-H, Jin C-B. Effects of a New Selective Phosphodiesterase Type 5 Inhibitor, KJH-1002, on the Relaxation of Rabbit Corpus Cavernosum Tissue. J Appl Pharm 2003; 11: 232-7.

[28]   Venhuis BJ, Barends DM, Zwaagstra ME, Kaste Dd. Recent Developments in Counterfeits and Imitations of Viagra, Cialis and Levitra: A 2005 - 2006 Update. The Netherlands: National Institute for Public Health and the Environment 2007; p. 49.

[29]   Instrumental Analysis Data of Illegal Compounds in Food. http://nifds.go.kr2013. NiFDS

[30]   Ge X, Li L, Koh H-L, Low M-Y. Identification of a new sildenafil analogue in a health supplement. J Pharm Biomed Anal 2011; 56(3): 491-6.
[http://dx.doi.org/10.1016/j.jpba.2011.06.004] [PMID: 21726974]

[31]   Wollein U, Eisenreich W, Schramek N. Identification of novel sildenafil-analogues in an adulterated herbal food supplement. J Pharm Biomed Anal 2011; 56(4): 705-12.
[http://dx.doi.org/10.1016/j.jpba.2011.07.012] [PMID: 21821375]

[32]   Li L, Low MY, Aliwarga F, *et al.* Isolation and identification of hydroxythiohomosildenafil in herbal dietary supplements sold as sexual performance enhancement products. Food Addit Contam Part A Chem Anal Control Expo Risk Assess 2009; 26(2): 145-51.
[http://dx.doi.org/10.1080/02652030802368757] [PMID: 19680883]

[33]   Oh SS-Y, Zou P, Low M-Y, Koh H-L. Detection of sildenafil analogues in herbal products for erectile dysfunction. J Toxicol Environ Health A 2006; 69(21): 1951-8.
[http://dx.doi.org/10.1080/15287390600751355] [PMID: 16982533]

[34]   Reepmeyer JC, d'Avignon DA. Structure elucidation of thioketone analogues of sildenafil detected as adulterants in herbal aphrodisiacs. J Pharm Biomed Anal 2009; 49(1): 145-50.
[http://dx.doi.org/10.1016/j.jpba.2008.10.007] [PMID: 19042103]

[35]   Low M-Y, Li L, Ge X, Kee C-L, Koh H-L. Isolation and structural elucidation of flibanserin as an adulterant in a health supplement used for female sexual performance enhancement. J Pharm Biomed Anal 2012; 57: 104-8.
[http://dx.doi.org/10.1016/j.jpba.2011.08.027] [PMID: 21955644]

[36]   Poplawska M, Blazewicz A, Zolek P, Fijalek Z. Determination of flibanserin and tadalafil in supplements for women sexual desire enhancement using high-performance liquid chromatography with tandem mass spectrometer, diode array detector and charged aerosol detector. J Pharm Biomed Anal 2014; 94(0): 45-53.
[http://dx.doi.org/10.1016/j.jpba.2014.01.021] [PMID: 24531007]

[37]   Cai Y, Cai T-G, Shi Y, *et al.* Simultaneous determination of eight PDE5-is potentially adulterated in herbal dietary supplements with TLC and HPLC-PDA-MS methods. J Liq Chromatogr Relat Technol 2010; 33(13): 1287-306.
[http://dx.doi.org/10.1080/10826076.2010.488979]

[38]   Moriyasu T, Shigeoka S, Kishimoto K, *et al.* [Identification system for Sildenafil in health foods].

Yakugaku Zasshi 2001; 121(10): 765-9.
[http://dx.doi.org/10.1248/yakushi.121.765] [PMID: 11676179]

[39]   Reddy TS, Reddy AS, Devi PS. Quantitative determination of sildenafil citrate in herbal medicinal formulations by high-performance thin-layer chromatography. *J. PLANAR.* CHROMAT 2006; 19(112): 427-31.

[40]   Do TT, Theocharis G, Reich E. Simultaneous Detection of Three Phosphodiesterase Type 5 Inhibitors and Eight of Their Analogs in Lifestyle Products and Screening for Adulterants by High-Performance Thin-Layer Chromatography. J AOAC Int 2015; 98(5): 1226-33.
[http://dx.doi.org/10.5740/jaoacint.14-285] [PMID: 26525240]

[41]   Zou P, Oh SS-Y, Hou P, Low M-Y, Koh H-L. Simultaneous determination of synthetic phosphodiesterase-5 inhibitors found in a dietary supplement and pre-mixed bulk powders for dietary supplements using high-performance liquid chromatography with diode array detection and liquid chromatography-electrospray ionization tandem mass spectrometry. J Chromatogr A 2006; 1104(1-2): 113-22.
[http://dx.doi.org/10.1016/j.chroma.2005.11.103] [PMID: 16364350]

[42]   Singh S, Prasad B, Savaliya AA, Shah RP, Gohil VM, Kaur A. Strategies for characterizing sildenafil, vardenafil, tadalafil, and their analogues in herbal dietary supplements, and detecting counterfeit products containing these drugs. Trends Analyt Chem 2008; 28(1): 13-28.
[http://dx.doi.org/10.1016/j.trac.2008.09.004]

[43]   Schramek N, Wollein U, Eisenreich W. Pyrazolopyrimidines in 'all-natural' products for erectile dysfunction treatment: the unreliable quality of dietary supplements. Food Addit Contam Part A Chem Anal Control Expo Risk Assess 2015; 32(2): 127-40.
[http://dx.doi.org/10.1080/19440049.2014.992980] [PMID: 25517174]

[44]   Mustazza C, Borioni A, Rodomonte AL, *et al.* Characterization of Sildenafil analogs by MS/MS and NMR: a guidance for detection and structure elucidation of phosphodiesterase-5 inhibitors. J Pharm Biomed Anal 2014; 96(0): 170-86.
[http://dx.doi.org/10.1016/j.jpba.2014.03.038] [PMID: 24747148]

[45]   Sabin GP, Lozano VA, Rocha WFC, Romão W, Ortiz RS, Poppi RJ. Characterization of sildenafil citrate tablets of different sources by near infrared chemical imaging and chemometric tools. J Pharm Biomed Anal 2013; 85(0): 207-12.
[http://dx.doi.org/10.1016/j.jpba.2013.07.036] [PMID: 23962562]

[46]   Mans DJ, Callahan RJ, Dunn JD, Gryniewicz-Ruzicka CM. Rapid-screening detection of acetildenafils, sildenafils and avanafil by ion mobility spectrometry. J Pharm Biomed Anal 2013; 75(0): 153-7.
[http://dx.doi.org/10.1016/j.jpba.2012.11.031] [PMID: 23262416]

[47]   Man CN, Nor NM, Lajis R, Harn GL. Identification of sildenafil, tadalafil and vardenafil by gas chromatography-mass spectrometry on short capillary column. J Chromatogr A 2009; 1216(47): 8426-30.
[http://dx.doi.org/10.1016/j.chroma.2009.10.016] [PMID: 19853256]

[48]   Guo J-B, Xu Y, Huang Z-B, He Q-H, Liu S-W. Development of an immunoassay for rapid screening of vardenafil and its potential analogues in herbal products based on a group specific monoclonal antibody. Anal Chim Acta 2010; 658(2): 197-203.
[http://dx.doi.org/10.1016/j.aca.2009.11.021] [PMID: 20103095]

[49]   Song Y, Wang YY, Zhang Y, Wang S. Development of enzyme-linked immunosorbent assay for rapid determination of sildenafil in adulterated functional foods. Food Agric Immunol 2011; 23(4): 338-51.
[http://dx.doi.org/10.1080/09540105.2011.630066]

[50]   Fejős I, Neumajer G, Béni S, Jankovics P. Qualitative and quantitative analysis of PDE-5 inhibitors in counterfeit medicines and dietary supplements by HPLC-UV using sildenafil as a sole reference. J Pharm Biomed Anal 2014; 98(0): 327-33.

[http://dx.doi.org/10.1016/j.jpba.2014.06.010] [PMID: 24996004]

[51]   Monakhova YB, Kuballa T, Löbell-Behrends S, *et al.* Standardless 1H NMR determination of pharmacologically active substances in dietary supplements and medicines that have been illegally traded over the internet. Drug Test Anal 2013; 5(6): 400-11.
[http://dx.doi.org/10.1002/dta.1367] [PMID: 22550015]

[52]   Venhuis BJ, Zwaagstra ME, Keizers PH, de Kaste D. Dose-to-dose variations with single packages of counterfeit medicines and adulterated dietary supplements as a potential source of false negatives and inaccurate health risk assessments. J Pharm Biomed Anal 2014; 89: 158-65.
[http://dx.doi.org/10.1016/j.jpba.2013.10.038] [PMID: 24291553]

[53]   Anonymous
http://www.hsa.gov.sg/content/hsa/en/News_Events/Press_Releases/2012/hsa_alerts_public0.html
(accessed 10.08.2015).

[54]   Guo B, Wang M, Liu Y, *et al.* Wide-Scope Screening of Illegal Adulterants in Dietary and Herbal Supplements *via* Rapid Polarity-Switching and Multistage Accurate Mass Confirmation Using an LC-IT/TOF Hybrid Instrument. J Agric Food Chem 2015; 63(31): 6954-67.
[http://dx.doi.org/10.1021/acs.jafc.5b02222] [PMID: 26189662]

[55]   Lv D, Cao Y, Lou Z, *et al.* Rapid on-site detection of ephedrine and its analogues used as adulterants in slimming dietary supplements by TLC-SERS. Anal Bioanal Chem 2015; 407(5): 1313-25.
[http://dx.doi.org/10.1007/s00216-014-8380-9] [PMID: 25542571]

[56]   Ma W, Ma Q, Fu L, *et al.* Simultaneous determination of rimonabant and orlistat illegally added in the weight-loss functional foods by HPLC-tandem mass spectrometry. Se Pu 2010; 28: 43-8.
[http://dx.doi.org/10.3724/SP.J.1123.2010.00043] [PMID: 20458919]

[57]   Chaves MA, Akatuka AS, Trujillo LM. Diethylpropion, fenproporex, diazepam and phenolphthalein - determination in weight-loss formulation. Rev Inst Adolfo Lutz 1994; 54: 36-43.

[58]   Qiu Y, Wang T, Li M, Pang X, Li J. Detection of phenolphthalein illegally mixed into slimming Chinese traditional medicines and health products by the liquid chromatography-quadrupole mass spectrometry method. Yaowu Fenxi Zazhi 2006; 26: 1609-11.

[59]   Qiu Y, Wu X-o, Feng F, Pang X, Li J. Identification of phenolphthalein, sibutramine and its two derivatives added illegally in slimming health care food. Zhongguo Yiyao Gongye Zazhi 2012; 43: 861-4.

[60]   Kim HJ, Lee JH, Park HJ, Cho SH, Cho S, Kim WS. Monitoring of 29 weight loss compounds in foods and dietary supplements by LC-MS/MS. Food Addit Contam Part A Chem Anal Control Expo Risk Assess 2014; 31(5): 777-83.
[http://dx.doi.org/10.1080/19440049.2014.888497] [PMID: 24499058]

[61]   Gad SC, Gad SE. Are dietary supplements safe as currently regulated? The great debate. Int J Toxicol 2003; 22(5): 381-5.
[http://dx.doi.org/10.1177/109158180302200508] [PMID: 14555411]

[62]   [FDA], Final rule declaring dietary supplements containing ephedrine alkaloids adulterated because they present an unreasonable risk. Final rule. Fed. Regist. 2004, 69 (28), 6787-854.

[63]   Vaysse J, Balayssac S, Gilard V, Desoubdzanne D, Malet-Martino M, Martino R. Analysis of adulterated herbal medicines and dietary supplements marketed for weight loss by DOSY 1H-NMR. Food Addit Contam Part A Chem Anal Control Expo Risk Assess 2010; 27(7): 903-16.
[http://dx.doi.org/10.1080/19440041003705821] [PMID: 20437283]

[64]   Yuen Y P, Lai C K, Poon W T, Ng S W, Chan A Y W, Mak T W L. Adulteration of over-the-counter slimming products with pharmaceutical analogue--an emerging threat. Hong Kong Med. J. 2007, 13 (Copyright (C) 2015 U.S. National Library of Medicine.), 216-20.

[65]   Chen SP, Tang MH, Ng SW, Poon WT, Chan AY, Mak TW. Psychosis associated with usage of herbal slimming products adulterated with sibutramine: a case series. Clin Toxicol (Phila) 2010; 48(8):

832-8.
[http://dx.doi.org/10.3109/15563650.2010.517208] [PMID: 20969504]

[66]   Mao H, Qi M, Zhou Y, *et al*. Discrimination of sibutramine and its analogues based on surface-enhanced Raman spectroscopy and chemometrics: toward the rapid detection of synthetic anorexic drugs in natural slimming products. RSC Advances 2015; 5: 5886-94.
[http://dx.doi.org/10.1039/C4RA09636C]

[67]   Jung J, Hermanns-Clausen M, Weinmann W. Anorectic sibutramine detected in a Chinese herbal drug for weight loss. Forensic Sci Int 2006; 161(2-3): 221-2.
[http://dx.doi.org/10.1016/j.forsciint.2006.02.052] [PMID: 16870382]

[68]   Anonymous Public Notification. http://www.fda.gov/Drugs/ResourcesForYou/Consumers/Buying UsingMedicineSafely/MedicationHealthFraud/ucm376059.htm2015.

[69]   Bertholee D, ter Horst PG, Wieringa A, Smit JP. [Life-threatening psychosis caused by using sibutramine-contaminated weight-loss coffee]. Ned Tijdschr Geneeskd 2013; 157(51): A6676. [Life-threatening psychosis caused by using sibutramine-contaminated weight-loss coffee].
[PMID: 24345358]

[70]   Connoley IP, Liu Y-L, Frost I, Reckless IP, Heal DJ, Stock MJ. Thermogenic effects of sibutramine and its metabolites. Br J Pharmacol 1999; 126(6): 1487-95.
[http://dx.doi.org/10.1038/sj.bjp.0702446] [PMID: 10217544]

[71]   [EMA], European Medicines Agency recommends suspension of marketing authorisations for sibutramine European Medicines Agency: London, 2010.

[72]   [FDA], Meridia (sibutramine): Marked withdrawal due to risk of serious cardiovascular events. U.S. Food and Drug Administration: 2010.

[73]   Kim JW, Kweon SJ, Park SK, *et al*. Isolation and identification of a sibutramine analogue adulterated in slimming dietary supplements. Food Addit Contam Part A Chem Anal Control Expo Risk Assess 2013; 30(7): 1221-9.
[http://dx.doi.org/10.1080/19440049.2013.793826] [PMID: 23799645]

[74]   Zou P, Oh SS-Y, Kiang K-H, Low M-Y, Bloodworth BC. Detection of sibutramine, its two metabolites and one analogue in a herbal product for weight loss by liquid chromatography triple quadrupole mass spectrometry and time-of-flight mass spectrometry. Rapid Commun Mass Spectrom 2007; 21(4): 614-8.
[http://dx.doi.org/10.1002/rcm.2876] [PMID: 17265544]

[75]   [HSA], HSA alerts public to two illegal products containing potent undeclared ingredients, including the banned chemical sibutramine. Singapore, 2014.

[76]   Mans DJ, Gucinski AC, Dunn JD, *et al*. Rapid screening and structural elucidation of a novel sibutramine analogue in a weight loss supplement: 11-desisobutyl-11-benzylsibutramine. J Pharm Biomed Anal 2013; 83: 122-8.
[http://dx.doi.org/10.1016/j.jpba.2013.02.031] [PMID: 23739298]

[77]   Huang F, Wu H-q, Huang X-l, *et al*. Simultaneous determination of six chemical drugs illegally added in dietary supplement by high performance liquid chromatography combined with tandem mass spectrometry. Fenxi Ceshi Xuebao 2013; 32: 699-704.

[78]   Rebiere H, Guinot P, Civade C, Bonnet PA, Nicolas A. Detection of hazardous weight-loss substances in adulterated slimming formulations using ultra-high-pressure liquid chromatography with diode-array detection. Food Addit Contam Part A Chem Anal Control Expo Risk Assess 2012; 29(2): 161-71.
[http://dx.doi.org/10.1080/19440049.2011.638676] [PMID: 22150438]

[79]   Reeuwijk NM, Venhuis BJ, de Kaste D, Hoogenboom RLAP, Rietjens IMCM, Martena MJ. Active pharmaceutical ingredients detected in herbal food supplements for weight loss sampled on the Dutch market. Food Addit Contam Part A Chem Anal Control Expo Risk Assess 2014; 31(11): 1783-93.

[http://dx.doi.org/10.1080/19440049.2014.958574] [PMID: 25247833]

[80]   Public statement on Acomplia (rimonabant) Withdrawl of the marketing authorization in the European Union. London: European Medicines Agency 2009.

[81]   Parascandola J. Dinitrophenol and bioenergetics: an historical perspective. Mol Cell Biochem 1974; 5(1-2): 69-77.
[http://dx.doi.org/10.1007/BF01874175] [PMID: 4610359]

[82]   Cutting WC, Mehrtens HG, Tainter ML. Actions and uses of dinitrophenol: Promising metabolic applications. J Am Med Assoc 1933; 101(3): 193-5.
[http://dx.doi.org/10.1001/jama.1933.02740280013006]

[83]   Cann HM, Verhulst HL. Fatality from acute dinitrophenol derivative poisoning. Am J Dis Child 1960; 100: 947-8.
[PMID: 13690446]

[84]   Le G, Dehaze C. Bull Soc Belge Ophtalmol 1959; 123: 499-504. [Two cases of cataract caused by dinitrophenol].
[PMID: 13853607]

[85]   Marigo M. [Medicolegal problems in dinitrophenol poisoning]. Riv Infort Mal Prof 1960; 47: 196-208. [Medicolegal problems in dinitrophenol poisoning].
[PMID: 13766747]

[86]   Chemistry CPa. Alpha-dinitrophenol, preliminary report of the Council on Pharmacy and Chemistry. Dinitrophenol not acceptable for New and Nonofficial Remedies. J Am Med Assoc 1933; 101(3): 208-10.

[87]   Chemistry CPa. Dinitrophenol not acceptable for New and nonofficial Remedies. J Am Med Assoc 1935; 105(1): 31-3.

[88]   Tainter ML, Cutting WC, Stockton AB. Use of Dinitrophenol in Nutritional Disorders : A Critical Survey of Clinical Results. Am J Public Health Nations Health 1934; 24(10): 1045-53.
[http://dx.doi.org/10.2105/AJPH.24.10.1045] [PMID: 18014064]

[89]   Le P, Wood B, Kumarasinghe SP. Cutaneous drug toxicity from 2,4-dinitrophenol (DNP): Case report and histological description. Australas J Dermatol 2014.
[PMID: 25367505]

[90]   Kurt TL, Anderson R, Petty C, Bost R, Reed G, Holland J. Dinitrophenol in weight loss: the poison center and public health safety. Vet Hum Toxicol 1986; 28(6): 574-5.
[PMID: 3788046]

[91]   Grundlingh J, Dargan PI, El-Zanfaly M, Wood DM. 2,4-dinitrophenol (DNP): a weight loss agent with significant acute toxicity and risk of death. J Med Toxicol 2011; 7(3): 205-12.
[http://dx.doi.org/10.1007/s13181-011-0162-6] [PMID: 21739343]

[92]   Anonymous P. Rep Carcinog 2011; 12: 342-4.
[PMID: 21863080]

[93]   [FDA], Laxative drug products for over-the-counter human use. Food and Drug Administration, HHS. Final rule. Fed. Regist. 1999,64 (19), 4535-40.

[94]   Cohen PA, Benner C, McCormick D. Use of a pharmaceutically adulterated dietary supplement, Pai You Guo, among Brazilian-born women in the United States. J Gen Intern Med 2012; 27(1): 51-6.
[http://dx.doi.org/10.1007/s11606-011-1828-0] [PMID: 21845487]

[95]   Moreira APL, Motta MJ, Dal Molin TR, Viana C, de Carvalho LM. Determination of diuretics and laxatives as adulterants in herbal formulations for weight loss. Food Addit Contam Part A Chem Anal Control Expo Risk Assess 2013; 30(7): 1230-7.
[http://dx.doi.org/10.1080/19440049.2013.800649] [PMID: 23782322]

[96]   Yamamoto S, Sumioka S, Fujioka M, Mikami E, Miyamoto K. Study on detection of drugs in

slimming health foods using GC-MS/MS. Shokuhin Eiseigaku Zasshi 2011; 52(6): 363-9.
[http://dx.doi.org/10.3358/shokueishi.52.363] [PMID: 22200804]

[97] Dunn JD, Gryniewicz-Ruzicka CM, Mans DJ, *et al.* Qualitative screening for adulterants in weight-loss supplements by ion mobility spectrometry. J Pharm Biomed Anal 2012; 71: 18-26.
[http://dx.doi.org/10.1016/j.jpba.2012.07.020] [PMID: 22902504]

[98] Hammadi R, Almardini MA. A fully validated HPLC-UV method for quantitative and qualitative determination of six adulterant drugs in natural slimming dietary supplements. Int. J. Pharm. Sci. Rev. Res. 2014,29, 171-174, 4 pp.

[99] Johansson M, Fransson D, Rundlöf T, Huynh N-H, Arvidsson T. A general analytical platform and strategy in search for illegal drugs. J. Pharm. Biomed. Anal. 2014, (0).
[http://dx.doi.org/10.1016/j.jpba.2014.07.026]

[100] Odoardi S, Castrignanò E, Martello S, Chiarotti M, Strano-Rossi S. Determination of anabolic agents in dietary supplements by liquid chromatography-high-resolution mass spectrometry. Food Addit Contam Part A Chem Anal Control Expo Risk Assess 2015; 32(5): 635-47.
[PMID: 25719897]

[101] Kamber M, Baume N, Saugy M, Rivier L. Nutritional supplements as a source for positive doping cases? Int J Sport Nutr Exerc Metab 2001; 11(2): 258-63.
[http://dx.doi.org/10.1123/ijsnem.11.2.258] [PMID: 11402257]

[102] Geyer H, Parr MK, Koehler K, Mareck U, Schänzer W, Thevis M. Nutritional supplements cross-contaminated and faked with doping substances. J Mass Spectrom 2008; 43(7): 892-902.
[http://dx.doi.org/10.1002/jms.1452] [PMID: 18563865]

[103] Geyer H, Parr MK, Mareck U, Reinhart U, Schrader Y, Schänzer W. Analysis of non-hormonal nutritional supplements for anabolic-androgenic steroids - results of an international study. Int J Sports Med 2004; 25(2): 124-9.
[http://dx.doi.org/10.1055/s-2004-819955] [PMID: 14986195]

[104] Abbate V, Kicman A T, Evans-Brown M, *et al.* Anabolic steroids detected in bodybuilding dietary supplements – a significant risk to public health. Drug Test. Anal. 2015, n/a-n/a.

[105] Rahnema CD, Crosnoe LE, Kim ED. Designer steroids - over-the-counter supplements and their androgenic component: review of an increasing problem. Andrology 2015; 3(2): 150-5.
[http://dx.doi.org/10.1111/andr.307] [PMID: 25684733]

[106] Bogusz MJ, Hassan H, Al-Enazi E, Ibrahim Z, Al-Tufail M. Application of LC-ESI-MS-MS for detection of synthetic adulterants in herbal remedies. J Pharm Biomed Anal 2006; 41(2): 554-64.
[http://dx.doi.org/10.1016/j.jpba.2005.12.015] [PMID: 16427237]

[107] Cho S-H, Park HJ, Lee JH, *et al.* Monitoring of 35 illegally added steroid compounds in foods and dietary supplements. Food Addit Contam Part A Chem Anal Control Expo Risk Assess 2014; 31(9): 1470-5.
[http://dx.doi.org/10.1080/19440049.2014.946100] [PMID: 25036882]

[108] Jiang S, Tan H, Guo C, Gong L, Shi F. Development of an ultra-high-performance liquid chromatography coupled to high-resolution quadrupole-Orbitrap mass spectrometry method for the rapid detection and confirmation of illegal adulterated sedative-hypnotics in dietary supplements. Food Addit. Contam. A 2015, Ahead of Print.

[109] Chen Y, Zhao L, Lu F, Yu Y, Chai Y, Wu Y. Determination of synthetic drugs used to adulterate botanical dietary supplements using QTRAP LC-MS/MS. Food Addit Contam Part A Chem Anal Control Expo Risk Assess 2009; 26(5): 595-603.
[http://dx.doi.org/10.1080/02652030802641880] [PMID: 19680934]

[110] Liang Q, Qu J, Luo G, Wang Y. Rapid and reliable determination of illegal adulterant in herbal medicines and dietary supplements by LC/MS/MS. J Pharm Biomed Anal 2006; 40(2): 305-11.
[http://dx.doi.org/10.1016/j.jpba.2005.07.035] [PMID: 16174560]

[111] https://lakemedelsverket.se/english/All-news/NYHETER-2013/Pharmaceutical-ingredients-i--one-out-of-three-eyelash-serums/2015. MPA

[112] Chen J, Lu W, Zhang Q, Jiang X. Determination of the active metabolite of sibutramine by liquid chromatography-electrospray ionization tandem mass spectrometry. J Chromatogr B Analyt Technol Biomed Life Sci 2003; 785(2): 197-203.
[http://dx.doi.org/10.1016/S1570-0232(02)00818-8] [PMID: 12554132]

[113] Trebilcock B. If Your Diet Drug Works http://www.prevention.com/mind-body/natural-remedies/diet-pill-dangers-truth-behind-weight-loss-supplements2015.

CHAPTER 8

Adulteration of Food Supplements and Analytical Methodologies for their Quality Control

L.O. Demirezer[1,*] and E. Ucakturk[2]

[1] *Hacettepe University, Faculty of Pharmacy, Department of Pharmacognosy, 06100 Ankara, Turkey*

[2] *Hacettepe University, Faculty of Pharmacy, Department of Analytical Chemistry, 06100 Ankara, Turkey*

Abstract: Medicinal plant extracts are complex mixtures containing many compounds. Their therapeutic properties may be due to the cumulative or synergistic effects of contained compounds. The adulteration or substitution of the herbal products especially food supplements is crucial problem in herbal products industry. Adulteration or substitution may be deliberate or accidental. Valuable herbal products are substituted with similar low quality products and used as adulterant which may or may not have any therapeutic potential as that of original product. The quality control studies are very important to provide active, high quality and safe food supplement. Therefore, fast, specific, accurate, inexpensive, and high throughput analytical methods are needed. Some chromatographic and spectroscopic techniques are used for identification and quantification of bioactive compounds, impurities, contaminants, and also other compounds present in the food supplements. The chromatographic fingerprint analysis offers significant advantages in analytical methods because of the complex and multi component matrix of the food supplements. Screening of the content and quantification of food supplements can be possible in a single fingerprint run.

Keywords: Adulteration, Food supplements, Quality control.

INTRODUCTION

Mankind has learned the therapeutic effects of various parts of plants: roots, leaves, flowers, fruits, *etc.* the trial-and-error methods or as a results of chance. Green wave trends has gained importance also in medicine, as in all fields. In this manner, the popularity of herbal medicines and/or food supplements (=dietary supplements) has increased.

---

* **Corresponding author L.O. Demirezer:** Hacettepe University, Faculty of Pharmacy, Dept. of Pharmacognosy, 06100 Ankara, Turkey; Tel: 903123051089/3053100; Fax: 903123114777; E-mail: omurd@hacettepe.edu.tr

Alankar Shrivastava (Ed.)

The reason that very high consumption of herbals is to convey messages of manufacturers about herbals which are free from the side effects and the most of people believe in this idea because of their long historical clinical practice [1].

The coalescence of herbal medicine into modern medicine is closely related to quality and safety-efficacy. Herbal extracts are licensed in various countries as traditional herbal medicine and some of them as food supplements. Food supplements are sold in many countries with little regulations, while there are strict regulations for pharmaceuticals [2].

Recently, it was reported that food supplements contain undeclared synthetic drugs [3,4]. This adulteration may cause unpredictable health problems or reduce the effect. Quality affects the efficacy and safety of herbal products. The quality of herbal products varies depending on internal, external and regulatory factors. Genetic diversity, daily and seasonal changes, the selection of the wrong species take place within the internal factors. They also affect the quality and quantity of the active chemical compounds of medicinal plants. The use of the right plant genus or species is only possible with the binominal identification of the plant by a specialist. Also an expert can tell that the plant has no medicinal value, or they may be even toxic. In some cases, similar plants or unwanted part of the plant would be used intentionally or unintentionally in herbal products. On the other hand, culture of plants, growing conditions such as soil type, geographical status (location, attitute *et al.*), climate, harvesting techniques, transportation, safekeeping, production technics, accidentally substitution and/or contamination, deliberate adulteration are external factors. There are different regulations in different countries for the registration of herbal products.

So let's take a brief look at what factors will affect the quality of botanical products;

Adulteration

As described by Miller *et al.* adulteration is, the addition of some components which are not a part of the normal component or removal of the active compound. Thus, the product value is reduced [5].

Adding low grade or wholly different drugs similar to that of original drugs. It is added partially or fully which is inferior or substandard in therapeutic and chemical properties. Certainly, low-quality source materials or finished products are therapeutically less effective. Adulterants show morphologically and therapeutically similarity to crude drugs but they are low quality thereby cheap.

Finished products can be adulterated with synthetic pharmaceutical agents and it can lead to adverse effects. *e.g.* Sibutramine or desmethyl sibutramine are added as adulterant to some dietary supplements for losing weight. Anti-inflammatory drugs (diclofenac, phenylbutazone, ibuprofen), antidiabetic drugs, sexual stimulants (sildenafil, tadalafil, thiosildenafil), glucocorticoids (dexamethasone, betamethasone, prednisolone, cortisone acetate, hydrocortisone), non-steroidal antihypertensive agents (amlodipine, valsartan, clonidine, metoprolol, chlorthiazide), and many similar synthetic compounds were reported in different studies as adulterants [6]. However, most are sold with prescription of these drugs and used under medical control and have some adverse effects. Unfortunately, these compounds are used in combination of traditional herbal medicines which are sold as a 100% natural product. These kinds of adulteration would cause serious health issues. Otherwise the relatively harmless fillers can be an issue if they do not appear on the label.

*Ginkgo biloba*extract contains quercetin, kaempferol, and isorhamnetin derivative flavonoids. United States and British Pharmacopeas indicate the tolerable flavonoids content in *Ginkgo biloba*supplements. To reduce the cost of *Ginkgo biloba*extract, some kinds of adulterations have been added. The compounds rutin or quercetin or extracts of *Kaempferia galanga*L. or*Sophora japonica* L. may be added to increase the content of the flavonoids [7,8]. In one of our previous study on quality control of *Ginkgo biloba* leaf extract, we encountered surprises on HPLC fingerprint analysis. We saw that *Ginkgo biloba* products licensed as food supplements are adulterer with rutin to reach the desired amount of total flavonoid glycosides, while herbal medicinal products were licensed by health authority have expected results [7].

Hoodia gordonii (Masson) Sweet ex Decne.is used as an appetite suppressant and is the only species known to be effective. Active marker of *Hoodia gordonii* is an oxypregnane steroidal glycoside. Other *Hoodia* species are thought to be used as adulterants [9, 10].

Laxative herbals are widely used for their slimming properties. For this purpose, the most widely used plants are *Cassia angustifolia* and *C. acutifolia*. In our previous study, we obtained interesting results. Although Sennoside B are responsible for the laxative effect of Senna leaves, we didn't determine Sennoside B in all tested herbal products [11]. The widely used adulterants are given in Table **1**.

Table 1. The widely used adulterants in herbal food supplements.

Herbal Food Supplements	Adulterants	Ref.
Ginkgo biloba supplements	Rutin, quercetin, and some flavonol glycoside, *Kaempferia galanga*, *Sophora japonica* extracts, *Ginkgo biloba* root bark	[8, 12]
Dietary food supplements promoted for sexual stimulant	yohimbine, sildenafil, vardenafil, tadalafil, homosildenafil, acetildenafil and hydroxyhomosildenafil	[13, 14]
Dietary food supplements promoted for weight loss.	sibutramine, fluoxetine or metronidazole, N-di-desmethylsibutramine; fenfluramine, phenolphthalein, N-mono-desmethyl sibutramine, and orlistat.	[15]
Dietary supplements promoted for hypoglycemic activity	Metformin, glibenclamide and rosiglitazone, gliclazide, phenformin, repaglinide	[16, 17]
Dietary supplements promoted for anti-hypertensive activity	amlodipine, indapamide and valsartan	[18]
Different food supplements	Caffeine (stimulant), furosemide (diuretic), norephedrine and ephedrine (stimulant, decongestant and anorexigen action)	[19]
Panax notoginseng food supplements (root of Panax notoginseng)	*Panax ginseng*(Asian ginseng) and *Panax quinquefolius* (American ginseng)	[20]
Chamomile extracts (*Chamomilla recutita*) food supplements	*Anthemis cotula*	[21]

Substitution

Intentional or accidental substitution can be made by using totally different drug species belonging to same family or with totally different drug or with other plant parts.

Stigma Croci, stigma of *Crocus sativus* L., that is effective on the bloodstream. *Carthamus tinctorius* L., *Hemerocallis fulva* (L.) L. and *Hemerocallis citrina* Baroni are sold in the market under the name Stigma Croci. However, C. sativus is the only plant of proven effectiveness. Others should be considered as adulterant or substituents [22].

The *Datura metel*L. and *Datura stramonium*L. can be given as an example. They contain tropane alkaloids therefore have effected as bronchodilator and inhibit respiratory mucous membrane secretion. Besides the alcoholic extract of *D. metel* shows anthelmentic activity. *D. metel* and *D. stramonium* are useful for the respiratory diseases, while as *D. metel* would be a better choice for anthelmentic application.

Two completely different plants, *Clerodendron indicum*(L.) Kuntzeand *Solanum virginianum* L. (Syn: *Solanum xanthocarpum* Shrad. & Wendl.) are substituted for their antihistamin activity [23]. Referring again to give an example, *Tribulus terrestris* L. and *Pedalium murex* L. can be substituted. Both also belong to different families which are Zygophyllaceae and Pedaliaceae respectively. Despite they have different chemical contents, the drug is frequently substituted. *Tribulus terrestris* contain alkaloid, rutin, chlorogenin, diosgenin butthere are flavonoids, ursolic acid, vanillin, sitosterol, and alkaloids in the *Pedalium murex* [24].

Different organs of the same or different plants accidentally or deliberately used interchangeably. For example, barks are used instead of roots or leaves are used instead of flowers. The bulb of *Eucomis autumnalis* is substituted with the rhizome of*Siphonochilus aethiopicus*, and the bark of*Ocotea bullata*is substituted with the bark of *Warburgia salutaris* [25]. The widely used substitution are given in Table **2**.

Table2. Substitutions made widely [23].

No.	Botanical name	Substitute drug Botanical name
1.	*Plumbagozeylanica*	*Baliospermummontanum*
2.	*Marsdeniatenacissima*	*Lanneacoromandelica*
3.	*Mimusopselengi*	*Nelumbonucifera*
4.	*Valerianawallichii*	*Saussrealappa*
5.	*Myristicafragrans*	*Syzigiumaromaticum*
		Myristicafragrans (fruits)
6.	*Inularacemosa*	*Saussrealappa*
		Ricinuscommunis (root)
7.	*Piperchaba*	*Piperlongum* (root)
8.	*Vitisvinifera*	*Gmelinaarborea*
9.	*Clerodendrumserratum*	*Solanumxanthocarpum*
10.	*Fagoniacretica*	*Alhagipseudalhagi*
11.	*Capparissepiaria*	*Alocasiaindica*
12.	*Mimusopselengi*	*Acaciaarabica* (bark
13.	*Ocimumsanctum*	*Vitexnegundo*
14.	*Hobenariaspp.*	*Dioscoreabulbifera*

(Table 2) contd.....

No.	Botanical name	Substitute drug Botanical name
15.	*Saccharumofficinarum*	*Arundodonax*
16.	*Liliumpolyphyllum*	*Withaniasomnifera*
17.	*Fritillariaroylei*	*Withaniasomnifera*
18.	*Semecarpusanacardium*	*Semecarpustravancorica*
19.	*Aconitumheterophyllum*	*Cyperusrotundus*
20.	*Punicagranatum*	*Garciniaindica*
21.	*Cinnamomumcamphora*	*Leonotisnepetafolia*
22.	*Mesuaferrea*	*Nelumbonucifera*
23.	*Desmostachyabipinnata*	*Saccharumspontaneum*
24.	*Ocimumbasilicum*	*Ocimumsanctum*
25.	*Garciniapedunculata*	*Garciniaindica*

Contamination

Plant materials, may be contaminated with impurities, heavy metals, pesticides, insecticides and environmental pollution.

In summary, adulteration changes the natural content of food supplements. All over the world there is a growing trend in the marketing of food supplements with adulterates. Uncontrolled quality of herbal products creates challenges for the modernization and internationalization. Regulatory guidelines and pharmacopoeias require macroscopic and microscopic assessment of the herbal material for quality control and standardization [26 - 28].

The regulation practices of the countries for herbal medicines are important. Differences in the quality of herbal products is due to the different regulations in various countries. To improve the quality of the products, the plants should be cultivated according to the rules of good agricultural practices (GAPs) and finished products must be manufactured according to good manufacturing practices (GMPs) guidelines and post-marketing products should be under observation. Their quality controls are necessary in the standardization process of them. Good Agricultural Practice (GAP), Good Agricultural and Collection Practice (GACP) and Good Manufacturing Practice (GMP) guidelines are discussed how the standardization of quality assurance and safety of manufacturing in food supplements. GAP contains production processes including seeding, cultivating, and raw material production. GMP guidelines consist of the steps from raw material to the consumption [29].

European Agency for the Evaluation of Medicinal Products, since 2002, has accepted the GAP guidelines for herbal products. In 2003, The World Health Organization (WHO) published good agricultural and collection practices (GACP) rules for medicinal plants. These guidelines contain some data including cultivation, collection, crop, post-crop processing, distribution, and safekeeping conditions of medicinal plants [30].

In United States, the Dietary Supplement Health and Education Act of 1994 described the herbs as food supplements. According to this legislation FDA regulates dietary supplements for Current Good Manufacturing Practices (CGMPs) under 21 CFR Part 111 and all manufacturing companies must comply with CGMPs. Also manufacturing companies must report any serious side effects associated with the use of the product. In the European Union, food supplements are sold specifying the potential indications and content, while in the United States are sold with little regulation unlike the drug [2].

One of the widely used quality control methods by industry is to analyze the chemical markers known to be present in herbal products. However, differences of geographical source, cultivation and processing methods can change the chemical composition and therefore the clinical efficacy. To achieve safe and high quality food and non-food agricultural products, environmental, economic and social sustainability are provided with Good Agricultural Practice [31]. A GMP (Good manufacturing practices) is a system in order to ensure that products are produced and controlled according to quality standards (ISPE). Good agricultural practice GAP and good manufacturing practice GMP are a set of principles, regulations and technical recommendations applicable to production and processing. Quality of herbal products and human health are closely related to GAP and GMP. Quality control of herbal products has not only to establish reasonable analytical methods for analysis of the active ingredients of herbal medicines, but also necessary to evaluate the many other factors, such as insecticides, herbicides and pesticides residue, heavy metals contamination and aflatoxin content.

Adulteration determination of herbal food supplements is challenge because of the complex nature of herbal matrix. Adulteration could be investigated by fingerprint analysis technique. This technique helps us to see similarities and differences between drugs and enables us to screen active markers, adulterants, contaminants, also other constitutes in herbal food supplements. In addition, marker compounds and adulterants could be quantified with fingerprint analysis if there are available reference standards.

Fingerprinting analysis technique is widely used for the authentication of herbal food supplements and it is globally accepted as quality control of herbal materials and their finished products. World Health Organization (WHO) recommend the use of chromatographic fingerprinting techniques, EMA (European Medicines Agency) and FDA (Food and Drug Administration) accept the chromatographic fingerprint technology for identification and quality estimation of herbal materials [32].

It is classified two categories:

- Spectral fingerprint techniques (such as ultraviolet spectrophotometry-UV, mass spectrometry - MS, nuclear magnetic resonance spectroscopy-NMR, Fourier transform infrared spectroscopy - FTIR) [33 - 35].

- Chromatographic fingerprint techniques (High performance thin layer chromatography-HPTLC, High performance liquid chromatography-HPLC, ultra-performance liquid chromatography-UPLC, gas chromatography GC, capillary electrophoresis CE).

Chromatographic fingerprint analysis of herbal medicines/food supplements perform a exhaustively qualitative and quantitative approach for authentication, and ensuring their consistency and stability.

Chromatographic fingerprint techniques include:

• Thin-layer chromatography (TLC)

• Liquid chromatography (LC and UPLC)

• Gas chromatography (GC)

It is noteworthy that fingerprint techniques mentioned above is named chemical fingerprint. Biological fingerprint techniques, which investigate the genetic composition of herbal product, have not considered in this chapter.

Determination of Adulterants by Chromatographic Methods

HPTLC is ideal analytical tool for rapid analysis of complex herbal mixtures and offers the ability to present the results as an image and adaptable single or multiple samples. Simplicity, cost of efficiency, parallel analysis of multiple sample, multiple detection options, high sample capacity, rapid results are the advantages of HPTLC. Although it has some disadvantages. The separation power of HPTLC is lower than that of HPLC and UPLC. There are many successful applications of HPTLC in the authentication of plant materials. For examples, 11 of pregnane glycosides in Hoodia products, and triterpene glycosides in

Cimicifuga (black cohosh) were successfully investigated by HPTLC [10, 36].

Herbal food supplements consist of different type of molecules. Reversed phase liquid chromatography (RP) is widely used for their chromatographic fingerprint analysis [37 - 42]. Nonpolar compounds could be analyzed by RP whereas Normal phase or Hydrophilic Interaction Chromatography (HILIC) is performed for polar and hydrophilic compounds. HILIC could be successfully performed for separation of these polar compounds in herbal food supplements such as *Ligusticum chuanxiong* Hort [42]. Two types of HILIC column (underivatized silica and amide-bonded silica) were tried for separation and retention of polar compounds in *Ligusticum chuanxiong*. Amide-bonded silica column was chosen because of the better retention and resolution of compounds. HILIC consumes more aqueous solvent than normal phase chromatography so it is more environmental friendly chromatographic techniques. Also aqueous organic solutions in mobile phase enhance the ionization of compounds in MS. One example for HILIC-MS application is Nucleobases and nucleosides determination in *Ginkgo biloba* leaves. Separation and quantitation of these polar compounds in *G. biloba* leaves were carried out by HILIC-UPLC-MS/MS [43].

Herbal food supplements are adulterated with some synthetic drugs. Metformin, synthetic anti-diabetic drug, is one of the most used adulterant. It is very polar molecule and its retention on RP column is poor. Therefore, some mobile phase additives such as ion pairing reagents are used in reversed phase chromatography. Ion pairing reagents help to retain and separation of polar compounds on RP column, but these reagents are not preferred for analysis because of long column equilibration time, and not suitable for MS detection. So HILIC is the best choice of metformin analysis in drug substances [44].

It is known that UPLC gives more separation efficiently and sensitivity, lower solvent consumption rate, shorter analysis time compared to conventional LC. Therefore, UPLC gives many advantages on fingerprint analysis. Thus, separation of many compounds in herbal food supplements is achieved in a short analysis time. For example, the separation of 12 of oxypregnane glycosides was implemented within 15 min in UPLC instead of the 80 min in conventional LC [9]. Many UPLC methods were reported for determination of adulterants in herbal medicine [45 - 47].

GC offers higher efficient, sensitive and fast chromatographic analysis. In GC-MS volatile and thermally stable molecules can be analyzed or the molecules unstable in high temperature are derivatized before analysis. Analysis of herbal food supplements in GC coupled with MS gives valuable knowledge such as unique fingerprint chromatogram from total ion chromatogram, also identification and quantification of the compounds would be possible. 68 Volatile compounds in *Teucrium chamaedrys* was performed by GC×GC–TOF/MS [48]. GC coupled with a flame ionization detector was used for determination of the amount of ephedrine-type alkaloids in herbs and commercial herbal products [49].

Besides above mentioned chromatographic analysis methods, capillary electrophoresis (CE) would prefer in many fingerprint analysis of plant extract [19, 50]. CE, with a small amount of sample has several advantages such as good and distinct separation and short analysis time. However, reproducibility between run is a limitation for CE. The most usable detection method in CE is either UV (or diode array) or MS. CE-UV analysis of phenolic compounds such as flavones, flavanols, flavanones, coumarins and phenylpropanoids of chamomile was carried out in less than 7.5 minutes [51].

Chromatographic Detection Methods of Adulterants in Herbal Medicine

Ultraviolet (UV), Diode array (DA), MS, NMR, evaporative light-scattering (ELS) detection methods are chosen according to the active markers, adulterants and other constitutes in herbal food supplements.

UV detection is mostly preferable because of its simplicity and cost effectiveness. Most of the natural compounds in herbal food supplements for example flavonoids, phenolic derivatives, quercetin, 3-*O*-glucosyl quercetin, and 3,4-di-*O*-glucosyl quercetin give absorption in UV-Visible region (200-780 nm). But some compounds such as triterpene glycosides, ginsenosides show weak UV absorption in near UV region. In this region solvents such as acetonitrile and methanol, and many organic compounds give absorption so selectivity of the chromatographic method is decreased. Therefore, instead of UV, ELS and MS detection could be performed.

DA detection is more valuable detection method comparing to UV. It enables us to record full UV–vis spectra of all compound in the matrix. Thus maximum absorption wavelength could be determined for compounds and many compounds having different UV characteristic could be analyzed in single LC run. In addition, DA detector allow us to check the purity of chromatographic peaks, which enable us to understand if there are any co-eluting compounds together with the interested compound. This feature is very important especially there lacks of opportunity to prepare a placebo sample for analysis like plant matrix.

ELS detection provides high sensitivity and baseline noise but it does not give any structural information like MS and NMR [52 - 54].

Quantitative and quantitative determination of adulterations could be performed using authentic standards and/or standardized plant extracts using UV/DA, ELS detection. But most of the time these standards/extract are commercially unavailable. Therefore, LC coupled with MS detection plays a critical role because of providing structural information and its high sensitivity, selectivity. It makes identification and quantification of nearly all ingredients (active compounds, adulterants and any other components) in plant matrix possible. Also LC coupled with high resolution mass spectrometry, tandem or multi-stage mass spectrometry give valuable information about unknown compounds in the mixture. These techniques provide unique fragmentation patterns and accurate mass so molecular formula of adulterants could be revealed [55]. Mass spectrometry libraries are also helpful for determination of synthetic undeclared and mislabeled adulterants such as synthetic PDE-5 inhibitor analogues, anti-hypertensive and steroidal and non-steroidal anti-inflammatory drugs [56, 57].

Ionization techniques, Electrospray Ionization (ESI), and Atmospheric Pressure Chemical Ionization (APCI), are used in LC-MS or LC-MS-MS. ESI is widely used techniques, but APCI is preferred for ionization of some compound such as qualification of isoflavonoids and astragalosides [58]. In addition, analysis of triterpene glycosides using LC combined with atmospheric pressure chemical ionization technique generates more fragment ions, thus more valuable structural information was obtained [59].

In the literature, more than one detection technique is preferred to use for adulteration determination which would be increased the reliability of the results. In most studies, fingerprint analysis in herbal food supplements is first done by UV detection and then structure of the adulterants and their analogs characterize in MS. In addition, as mentioned before, herbal food supplements contain different kinds of molecules so that different separation and detection techniques are necessary. One of the examples is *Ginkgo biloba* preparations licensed as food supplements. These preparations have quercetin, kaempferol, and isorhamnetin derivative different flavonoids and also terpenoids. Analysis of flavonoids in these preparations is mostly performed by LC-UV or DA detection whereas LC coupled with MS detection is used for determination of their terpenoids content because of their poor UV absorption [7].

In recent years, NMR is used for detector for LC systems. Using NMR give us full structural information without need of reference standards/standardized extract or MS library. However, its sensitivity is lower than the other LC detectors [60]. Determination of adulterants in herbal food supplements promoted for anti-inflammatory and analgesic properties was done by LC–UV–SPE–NMR [61]. LC–MS–SPE/NMR is also used for adulterants in a Chinese herbal medicine [61].

Validation of Chromatographic Methods

Chromatographic fingerprint methods should be validated in order to verify the efficiency of the analytical method.

Validation studies of herbal food supplements would be investigated in two parts:

• Methods investigate the quantity of active markers in the plant

• Chromatographic fingerprint method

Analytical methods determined active markers in herbal food supplements could be validated ICH guideline [62]. The following validation parameters should be investigated: specificity, precision (inter- and intra-day precision, repeatability), linearity, limit of detection and quantification, accuracy (inter- and intra-day accuracy), stability, robustness and ruggedness. There may be some difficulties for validation studies. One of them is unavailable reference standard of active markers. Reference standards are obtained either extraction from plant matrix or synthesized by chemically. In addition, preparation of the placebo plant sample is hard and sometimes impossible. Standard addition method is very helpful to show the accuracy, and also specificity of analytical method especially if matrix is unknown and not obtained.

The validation parameters such as precision (inter- and intra-day), injection repeatability, stability, the robustness and ruggedness should be considered if the method is a chromatographic fingerprint method.

Internal standard is necessary to improve precision of analytical method because of the complex matrix and multiple compounds to be analyzed. One or sometimes more than one internal standard could be needed in case multiple active markers having different chemical characteristics.

CONCLUDING REMARKS

Traditional herbal medicines are derived from medicinal plants that contain many bioactive therapeutic molecules and compared to synthetic drugs it is known to have less side effects. Nevertheless, as mentioned above worldwide there is a rising trend in the marketing of food supplements or natural products that contain harmful chemicals and hidden synthetic drugs.

Quality controlled traditional herbal medicinal products and uncontrolled food supplements should be considered as dissimilar. In many countries there are no specific rules for the control of food supplements. Particularly slimming, sexual dysfunction improvement and body building products are sold as 100% natural products. Therefore, consumers are misled.

The pharmacological activity of a herbal product is associated with its Phytoconstituents. That means, adulteration changes the quality, safety and efficacy of herbal products. Moreover, by making false claims about herbal products, uncertified sellers are not only harming a large population of consumers, but are also giving a bad name to herbal medicine in general.

Herbal products like synthetic medicines quality, efficacy and safety criteria must provide. Different official books, which are Pharmacopeas and various regulatory guidelines including W.H.O guidelines to produce the herbal products should be considered. Furthermore, the contamination of herbal products with pesticides, insecticides and fumigants should be tested.

Different chromatographic and detection methods require for the fingerprint analysis because of the complex nature of plant materials. Development of the fully validated analytical fingerprint methods is needed for evaluating the quality of the products The fingerprint analysis is to be standardized for each herbal dietary food supplements. Thus, this standardization helps to construct the fingerprint database of the herbal dietary food supplements.

In conclusion, food supplements should be under strict control of the concerned authorities of the countries. The quality of those products should be pursued orderly. Society should be informed by the authorities about such products through the media. Users should consult a physician before using any food supplement or herbal medicine.

Of course, traditional herbal medicines have had a positive role in the treatment of diseases and have long and successful history. Therefore, it is not necessary to add synthetic chemicals to increase the effect. The manufacturers must avoid such kind of bad practices for endangering the human health and the value of herbal medicines should not be reduced.

CONSENT FOR PUBLICATION

Not applicable.

CONFLICT OF INTEREST

The author (editor) declares no conflict of interest, financial or otherwise.

ACKNOWLEDGNEMT

Declared none.

REFERENCES

[1]   Marcus DM, Grollman AP. Botanical medicines--the need for new regulations. N Engl J Med 2002; 347(25): 2073-6.
        [http://dx.doi.org/10.1056/NEJMsb022858] [PMID: 12490692]

[2]   Halsted CH. Dietary supplements and functional foods: 2 sides of a coin? Am J Clin Nutr 2003; 77(4) (Suppl.): 1001S-7S.
        [http://dx.doi.org/10.1093/ajcn/77.4.1001S] [PMID: 12663307]

[3]   Sarker MR. Adulteration of herbal medicines and dietary supplements with undeclared synthetic drugs: dangerous for human health. Int J Pharm Pharm Sci 2014; 6(4): 1-2.

[4]   Csupor D, Boros K, Dankó B, Veres K, Szendrei K, Hohmann J. Rapid identification of sibutramine in dietary supplements using a stepwise approach. Pharmazie 2013; 68(1): 15-8.
        [PMID: 23444775]

[5]   Miller LG, Hume A, Harris IM, *et al.* Adulteration: its various meanings. Pharmacotherapy 2001; 21(6): 770-1.
        [http://dx.doi.org/10.1592/phco.21.7.770.34576] [PMID: 11401191]

[6]   Vaclavik L, Krynitsky AJ, Rader JI. Mass spectrometric analysis of pharmaceutical adulterants in products labeled as botanical dietary supplements or herbal remedies: a review. Anal Bioanal Chem 2014; 406(27): 6767-90.
        [http://dx.doi.org/10.1007/s00216-014-8159-z] [PMID: 25270866]

[7]   Demirezer LÖ, Büyükkaya A, Ucakturk E, *et al.* Adulteration Determining of Pharmaceutical Forms of Ginkgo biloba Extracts from Different International Manufacturers. Rec Nat Prod 2014; 8(4): 394-400.

[8]   Harnly JM, Luthria D, Chen P. Detection of adulterated Ginkgo biloba supplements using chromatographic and spectral fingerprints. J AOAC Int 2012; 95(6): 1579-87.
        [http://dx.doi.org/10.5740/jaoacint.12-096] [PMID: 23451372]

[9]   Avula B, Wang YH, Pawar RS, Shukla YJ, Smillie TJ, Khan IA. A rapid method for chemical fingerprint analysis of Hoodia species, related genera, and dietary supplements using UPLC-UV-MS. J Pharm Biomed Anal 2008; 48(3): 722-31.
        [http://dx.doi.org/10.1016/j.jpba.2008.07.005] [PMID: 18718731]

[10]   Rumalla CS, Avula B, Shukla YJ, *et al.* Chemical fingerprint of Hoodia species, dietary supplements, and related genera by using HPTLC. J Sep Sci 2008; 31(22): 3959-64.
        [http://dx.doi.org/10.1002/jssc.200800441] [PMID: 19065611]

[11]   Demirezer LO, Karahan N, Ucakturk E, *et al.* HPLC Fingerprinting of Sennosides in Laxative Drugs with Isolation of Standard Substances from Some Senna Leaves. Rec Nat Prod 2011; 5(4): 261-70.

[12]   Wohlmuth H, Savage K, Dowell A, Mouatt P. Adulteration of Ginkgo biloba products and a simple method to improve its detection. Phytomedicine 2014; 21(6): 912-8.
        [http://dx.doi.org/10.1016/j.phymed.2014.01.010] [PMID: 24566389]

[13]   Shi F, Guo C, Gong L, *et al.* Application of a high resolution benchtop quadrupole-Orbitrap mass spectrometry for the rapid screening, confirmation and quantification of illegal adulterated phosphodiesterase-5 inhibitors in herbal medicines and dietary supplements. J Chromatogr A 2014; 1344: 91-8.
        [http://dx.doi.org/10.1016/j.chroma.2013.12.030] [PMID: 24377735]

[14]   Zou P, Oh SS, Hou P, Low MY, Koh HL. Simultaneous determination of synthetic phosphodiesterase-5 inhibitors found in a dietary supplement and pre-mixed bulk powders for dietary supplements using high-performance liquid chromatography with diode array detection and liquid chromatography-electrospray ionization tandem mass spectrometry. J Chromatogr A 2006; 1104(1-2): 113-22.
        [http://dx.doi.org/10.1016/j.chroma.2005.11.103] [PMID: 16364350]

[15] Wang J, Chen B, Yao S. Analysis of six synthetic adulterants in herbal weight-reducing dietary supplements by LC electrospray ionization-MS. Food Additives contaminants: part A 2008; 25(7): 822-30.
[http://dx.doi.org/10.1080/02652030801946553]

[16] Guo C, Shi F, Jiang S, *et al.* Simultaneous identification, confirmation and quantitation of illegal adulterated antidiabetics in herbal medicines and dietary supplements using high-resolution benchtop quadrupole-Orbitrap mass spectrometry. J Chromatogr B Analyt Technol Biomed Life Sci 2014; 967: 174-82.
[http://dx.doi.org/10.1016/j.jchromb.2014.07.032] [PMID: 25108366]

[17] Zhu Q, Cao Y, Cao Y, Chai Y, Lu F. Rapid on-site TLC-SERS detection of four antidiabetes drugs used as adulterants in botanical dietary supplements. Anal Bioanal Chem 2014; 406(7): 1877-84.
[http://dx.doi.org/10.1007/s00216-013-7605-7] [PMID: 24452744]

[18] Kesting JR, Huang J, Sørensen D. Identification of adulterants in a Chinese herbal medicine by LC-HRMS and LC-MS-SPE/NMR and comparative *in vivo* study with standards in a hypertensive rat model. J Pharm Biomed Anal 2010; 51(3): 705-11.
[http://dx.doi.org/10.1016/j.jpba.2009.09.043] [PMID: 19850434]

[19] Cianchino V, Acosta G, Ortega C, Martínez LD, Gomez MR. Analysis of potential adulteration in herbal medicines and dietary supplements for the weight control by capillary electrophoresis. Food Chem 2008; 108(3): 1075-81.
[http://dx.doi.org/10.1016/j.foodchem.2007.11.042] [PMID: 26065773]

[20] Wang CZ, Ni M, Sun S, *et al.* Detection of adulteration of notoginseng root extract with other panax species by quantitative HPLC coupled with PCA. J Agric Food Chem 2009; 57(6): 2363-7.
[http://dx.doi.org/10.1021/jf803320d] [PMID: 19256509]

[21] Guzelmeric E, Vovk I, Yesilada E. Development and validation of an HPTLC method for apigenin 7-O-glucoside in chamomile flowers and its application for fingerprint discrimination of chamomile-like materials. J Pharm Biomed Anal 2015; 107: 108-18.
[http://dx.doi.org/10.1016/j.jpba.2014.12.021] [PMID: 25575175]

[22] Ma XQ, Zhu DY, Li SP, Dong TT, Tsim KW. Authentic identification of stigma Croci (stigma of Crocus sativus) from its adulterants by molecular genetic analysis. Planta Med 2001; 67(2): 183-6.
[http://dx.doi.org/10.1055/s-2001-11533] [PMID: 11301875]

[23] Prakash O, Jyoti Kumar A, Kumar P, Kumar Manna N. Adulteration and Substitution in Indian Medicinal Plants: An Overview. J Med Plants Studies 2013; 1(4): 127-32.

[24] Kevalia J, Patel B. Identification of fruits of Tribulus terrestris Linn. and Pedalium murex Linn.: A pharmacognostical approach. Ayu 2011; 32(4): 550-3.
[http://dx.doi.org/10.4103/0974-8520.96132] [PMID: 22661853]

[25] Zschocke S, Rabe T, Taylor JLS, Jäger AK, van Staden J. Plant part substitution--a way to conserve endangered medicinal plants? J Ethnopharmacol 2000; 71(1-2): 281-92.
[http://dx.doi.org/10.1016/S0378-8741(00)00186-0] [PMID: 10904175]

[26] Quality Control Methods for Medicinal Plant Materials. Geneva: WHO 1998.

[27] Indian Herbal Pharmacopoeia. Mumbai: Indian Drug Manufacturers' Association 2002.

[28] British Herbal Pharmacopoeia. British Herbal Medicine Association 1996.

[29] Zhang AL, Xue CC, Fong HHS. Integration of Herbal Medicine into Evidence-Based Clinical Practice: Current status and Issues.Herbal Medicine: Biomolecular and Clinical Aspects. 2nd ed. Crc Press/Taylor and Francis 2011; pp. 1-18.
[http://dx.doi.org/10.1201/b10787-23]

[30] Fu PP, Chiang HM, Xia Q, *et al.* Quality assurance and safety of herbal dietary supplements. J Environ Sci Health C Environ Carcinog Ecotoxicol Rev 2009; 27(2): 91-119.

[http://dx.doi.org/10.1080/10590500902885676] [PMID: 19412857]

[31] Development of a Framework for Good Agricultural Practices. Seventeenth Session. 31 March-4 April 2003.

[32] Silva MLS. Comprehensive Analysis of Phytopharmaceutical Formulations–An Emphasis on Two-Dimensional Liquid Chromatography. J Chromatogr Sep Tech 2015; 6(2): 1-9.

[33] Black C, Haughey SA, Chevallier OP, Galvin-King P, Elliott CT. A comprehensive strategy to detect the fraudulent adulteration of herbs: The oregano approach. Food Chem 2016; 210: 551-7.
[http://dx.doi.org/10.1016/j.foodchem.2016.05.004] [PMID: 27211681]

[34] Zhang J, Zhang X, Dediu L, Victor C. Review of the current application of fingerprinting allowing detection of food adulteration and fraud in China. Food Control 2011; 22: 1126-35.
[http://dx.doi.org/10.1016/j.foodcont.2011.01.019]

[35] Singh SK, Jha SK, Chaudhary A, Yadava RDS, Rai SB. Quality control of herbal medicines by using spectroscopic techniques and multivariate statistical analysis. Pharm bio 2009; 48 (2): 134-41.

[36] Avula B, Wang YH, Rumalla CS, Ali Z, Smillie TJ, Khan IA. Analytical methods for determination of magnoflorine and saponins from roots of Caulophyllum thalictroides (L.) Michx. using UPLC, HPLC and HPTLC. J Pharm Biomed Anal 2011; 56(5): 895-903.
[http://dx.doi.org/10.1016/j.jpba.2011.07.028] [PMID: 21872415]

[37] Lu YL, Zhou NL, Liao SY, et al. Detection of adulteration of anti-hypertension dietary supplements and traditional Chinese medicines with synthetic drugs using LC/MS. Food Addit Contam Part A Chem Anal Control Expo Risk Assess 2010; 27(7): 893-902.
[http://dx.doi.org/10.1080/19440040903426710] [PMID: 20544454]

[38] Wang J, Chen B, Yao S. Analysis of six synthetic adulterants in herbal weight-reducing dietary supplements by LC electrospray ionization-MS. Food Addit Contam Part A Chem Anal Control Expo Risk Assess 2008; 25(7): 822-30.
[http://dx.doi.org/10.1080/02652030801946553] [PMID: 18569001]

[39] Yang J, Wang AQ, Li XJ, Fan X, Yin SS, Lan K. A chemical profiling strategy for semi-quantitative analysis of flavonoids in Ginkgo extracts. J Pharm Biomed Anal 2016; 123: 147-54.
[http://dx.doi.org/10.1016/j.jpba.2016.02.017] [PMID: 26907698]

[40] Ulloa J, Sambrotta L, Redko F, et al. Detection of a tadalafil analogue as an adulterant in a dietary supplement for erectile dysfunction. J Sex Med 2015; 12(1): 152-7.
[http://dx.doi.org/10.1111/jsm.12759] [PMID: 25402198]

[41] Yu C, Wang CZ, Zhou CJ, et al. Adulteration and cultivation region identification of American ginseng using HPLC coupled with multivariate analysis. J Pharm Biomed Anal 2014; 99: 8-15.
[http://dx.doi.org/10.1016/j.jpba.2014.06.031] [PMID: 25044150]

[42] Jin Y, Liang T, Fu Q, et al. Fingerprint analysis of Ligusticum chuanxiong using hydrophilic interaction chromatography and reversed-phase liquid chromatography. J Chromatogr A 2009; 1216(11): 2136-41.
[http://dx.doi.org/10.1016/j.chroma.2008.04.010] [PMID: 18440542]

[43] Yao X, Zhou G, Tang Y, Guo S, Qian D, Duan JA. HILIC-UPLC-MS/MS combined with hierarchical clustering analysis to rapidly analyze and evaluate nucleobases and nucleosides in Ginkgo biloba leaves. Drug Test Anal 2015; 7(2): 150-7.
[http://dx.doi.org/10.1002/dta.1634] [PMID: 24665003]

[44] Wu X, Zhu B, Lu L, Huang W, Pang D. Optimization of a solid phase extraction and hydrophilic interaction liquid chromatography-tandem mass spectrometry method for the determination of metformin in dietary supplements and herbal medicines. Food Chem 2012; 133(2): 482-8.
[http://dx.doi.org/10.1016/j.foodchem.2012.01.005] [PMID: 25683423]

[45] Avula B, Sagi S, Gafner S, et al. Identification of Ginkgo biloba supplements adulteration using high performance thin layer chromatography and ultra high performance liquid chromatography-diode array

detector-quadrupole time of flight-mass spectrometry. Anal Bioanal Chem 2015; 407(25): 7733-46.
[http://dx.doi.org/10.1007/s00216-015-8938-1] [PMID: 26297458]

[46]   Guo C, Shi F, Jiang S, *et al.* Simultaneous identification, confirmation and quantitation of illegal
        adulterated antidiabetics in herbal medicines and dietary supplements using high-resolution benchtop
        quadrupole-Orbitrap mass spectrometry. J Chromatogr B Analyt Technol Biomed Life Sci 2014; 967:
        174-82.
        [http://dx.doi.org/10.1016/j.jchromb.2014.07.032] [PMID: 25108366]

[47]   Li N, Cui M, Lu X, Qin F, Jiang K, Li F. A rapid and reliable UPLC-MS/MS method for the
        identification and quantification of fourteen synthetic anti-diabetic drugs in adulterated Chinese
        proprietary medicines and dietary supplements. Biomed Chromatogr 2010; 24(11): 1255-61.
        [http://dx.doi.org/10.1002/bmc.1438] [PMID: 20954219]

[48]   Ozel MZ, Göğüş F, Lewis AC. Determination of Teucrium chamaedrys volatiles by using direct
        thermal desorption-comprehensive two-dimensional gas chromatography-time-of-flight mass
        spectrometry. J Chromatogr A 2006; 1114(1): 164-9.
        [http://dx.doi.org/10.1016/j.chroma.2006.02.036] [PMID: 16516906]

[49]   Wang M, Marriott PJ, Chan WH, Lee AWM, Huie CW. Enantiomeric separation and quantification of
        ephedrine-type alkaloids in herbal materials by comprehensive two-dimensional gas chromatography.
        J Chromatogr A 2006; 1112(1-2): 361-8.
        [http://dx.doi.org/10.1016/j.chroma.2005.12.043] [PMID: 16387317]

[50]   Cianchino V, Ortega C, Acosta G, Martínez LD, Gomez MR. Fingerprint analysis and synthetic
        adulterant search in Hedera helix formulations by capillary electrophoresis. Pharmazie 2007; 62(4):
        262-5.
        [PMID: 17484280]

[51]   Fonseca FN, Tavares MFM, Horváth C. Capillary electrochromatography of selected phenolic
        compounds of Chamomilla recutita. J Chromatogr A 2007; 1154(1-2): 390-9.
        [http://dx.doi.org/10.1016/j.chroma.2007.03.106] [PMID: 17459397]

[52]   Chen Z, Zhang M, Li X, Xu Q, Yang S. [Simultaneous determination of five glycosides in Aidi
        injection by RP-HPLC-ELSD]. Zhongguo Zhongyao Zazhi 2011; 36(6): 706-8.
        [PMID: 21710733]

[53]   Wang CZ, Ni M, Sun S, *et al.* Detection of adulteration of notoginseng root extract with other *panax*
        species by quantitative HPLC coupled with PCA. J Agric Food Chem 2009; 57(6): 2363-7.
        [http://dx.doi.org/10.1021/jf803320d] [PMID: 19256509]

[54]   Deconinck E, De Leersnijder C, Custers D, Courselle P, De Beer JO. A strategy for the identification
        of plants in illegal pharmaceutical preparations and food supplements using chromatographic
        fingerprints. Anal Bioanal Chem 2013; 405(7): 2341-52.
        [http://dx.doi.org/10.1007/s00216-012-6649-4] [PMID: 23307125]

[55]   Patel DN, Li L, Kee CL, Ge X, Low MY, Koh HL. Screening of synthetic PDE-5 inhibitors and their
        analogues as adulterants: analytical techniques and challenges. J Pharm Biomed Anal 2014; 87: 176-
        90.
        [http://dx.doi.org/10.1016/j.jpba.2013.04.037] [PMID: 23721687]

[56]   Bogusz MJ, Hassan H, Al-Enazi E, Ibrahim Z, Al-Tufail M. Application of LC-ESI-MS-MS for
        detection of synthetic adulterants in herbal remedies. J Pharm Biomed Anal 2006; 41(2): 554-64.
        [http://dx.doi.org/10.1016/j.jpba.2005.12.015] [PMID: 16427237]

[57]   Savaliya AA, Shah RP, Prasad B, Singh S. Screening of Indian aphrodisiac ayurvedic/herbal
        healthcare products for adulteration with sildenafil, tadalafil and/or vardenafil using LC/PDA and
        extracted ion LC-MS/TOF. J Pharm Biomed Anal 2010; 52(3): 406-9.
        [http://dx.doi.org/10.1016/j.jpba.2009.05.021] [PMID: 19540696]

[58]   Zhang X, Xiao HB, Xue XY, Sun YG, Liang XM. Simultaneous characterization of isoflavonoids and
        astragalosides in two *Astragalus* species by high-performance liquid chromatography coupled with

atmospheric pressure chemical ionization tandem mass spectrometry. J Sep Sci 2007; 30(13): 2059-69.
[http://dx.doi.org/10.1002/jssc.200700014] [PMID: 17657828]

[59]   He K, Zheng B, Kim CH, Rogers L, Zheng Q. Direct analysis and identification of triterpene glycosides by LC/MS in black cohosh, Cimicifuga racemosa, and in several commercially available black cohosh products. Planta Med 2000; 66(7): 635-40.
[http://dx.doi.org/10.1055/s-2000-8619] [PMID: 11105569]

[60]   Silva LMA, Filho EGA, Thomasi SS, Silva BF, Ferreira AG, Venâncio T. Use of diffusion-ordered NMR spectroscopy and HPLC-UV-SPE-NMR to identify undeclared synthetic drugs in medicines illegally sold as phytotherapies. Magn Reson Chem 2013; 51(9): 541-8.
[http://dx.doi.org/10.1002/mrc.3984] [PMID: 23818305]

[61]   Kesting JR, Huang J, Sørensen D. Identification of adulterants in a Chinese herbal medicine by LC-HRMS and LC-MS-SPE/NMR and comparative *in vivo* study with standards in a hypertensive rat model. J Pharm Biomed Anal 2010; 51(3): 705-11.
[http://dx.doi.org/10.1016/j.jpba.2009.09.043] [PMID: 19850434]

[62]   International Conference of Harmonisation (ICH) Harmonised Tripartite Guideline—Validation of Analytical Procedures: Methodology Q2B. November 1996.

CHAPTER 9

Brief Overview of *Trans* Fat Analysis in Some Foods

Syed Tufail Hussain Sherazi[1,*], Sarfaraz Ahmed Mahesar[1], Aftab Ahmed Kandhro[2] and Sirajuddin[1]

[1] *National Centre of Excellence in Analytical Chemistry, University of Sindh, Jamshoro-76080, Pakistan*

[2] *Dr. M.A. Kazi Institute of Chemistry, University of Sindh, Jamshoro-76080, Pakistan*

Abstract: There are three macro components of foods namely protein, carbohydrates and lipids. Lipids include fats and oils which are very important components for the human diet and considered to be concentrated sources of energy. Major classes of fats and oils are vegetables oils, marine oils and animal fats. Naturally, in vegetable oils all fatty acids are in cis form. But through industrial processing, especially during hydrogenation process, artificial *trans* fats are developed and considered to be very dangerous for human health. According to the Food & Drug Administration (FDA), *trans* fat is supposed to be unwanted fat and in the diet it should be as minimum as possible. Therefore, *trans* fat labeling is mandatory in all foods in which fats and oils are involved. For the *trans* fat, labeling on food products requires adequate analytical methods. Therefore, the present review will cover brief outline of analytical techniques used for the analysis of *trans* fat.

Keywords: Analytical techniques, Food, Oils and fats, *Trans* fat.

INTRODUCTION

The characteristics of oils and fats depend upon the type and proportion of fatty acids (FAs). The nutritional performance of FAs is greatly influenced by the chain length of carbons, presence of double bonds, location of double bonds, cis-*trans* isomers and position of the FAs on the glycerol molecule [1, 2]. Each type of fatty acid has its own role on the health of consumers. Numerous studies have been conducted on the positive and negative effects of FAs present in the fats and oils [3].

---

* **Corresponding author Syed Tufail Hussain Sherazi:** National Centre of Excellence in Analytical Chemistry, University of Sindh, Jamshoro-76080, Pakistan; Tel: +92-229213429; Fax: +92-229213430; E-mail: tufail.sherazi@yahoo.com

Alankar Shrivastava (Ed.)

SATURATED FATTY ACIDS (SFAS)

According to many reported studies, an association of SFAs intake is linked with incident of coronary heart diseases (CHD) [3 - 6] but few studies quoted that intake of high SFAs not significantly influences the increased risk of CHD [7]. Saturated fat has been considered as a potential risk with increased level of low density lipoprotein (LDL) cholesterol [8]. Influence of dietary fat quality on some aspects of a moderate substitution of SFAs with MUFA on lipid metabolism in healthy persons was found to be beneficial [9, 10]. According to another reported study 22.37% cardiovascular disease (CVD) can be reduced by the daily energy intake of saturated and unsaturated fats and carbohydrates up to 5% [11].

Effects of certain parameters such as dietary intake, biochemical and anthropometriconobese children with and without fatty liver (FL) increased proportionally to the degree of hepatic steatosis due to intake of high SFAs [12]. Similarly, long chain SFAs provoked pro-inflammatory responses and significantly influenced on the growth and viability of endothelial cells (EC), while rising dietary intake of SFAs also shows association with increased inflammatory markers [13, 14].

MONOUNSATURATED FATTY ACIDS (MUFAS)

On the basis of double bonds MUFAs are differentiated from other fatty acid classes, as these contain only 1 double bond as compared to polyunsaturated fatty acids (PUFAs) with 2 or more double bonds. *Trans*-configuration contains hydrogen atoms on the opposite side of the double bond, whereas cis-configuration contains hydrogen atoms on same sides. In food materials the predominant form of MUFA isomers are in cis-configuration. In daily nutrition, oleic acid is the most common cis-configured MUFA available in the diet (~90% of all MUFAs) followed by palmitoleic acid and vaccenic acid [15, 16].

MUFAs are the most valuable in the diet when in *cis* form, because it facilitates to reduce the bad cholesterol and increase the good cholesterol high density lipoprotein (HDL) [17, 18]. A clinical study was conducted on the safety evaluation of medium chain fatty acids (MCFAs) and long chain fatty acids (LCFAs). Small intestine absorbs MCFA and transfers to the liver through bloodstream for hepatic metabolism, whereas LCFAs enter the lymphatic system through chylomicrons [19].

According to some reported studies, the medium chain triglyceride (MCT) and proper omega-6/omega-3 ratio may decrease dietary lipid storage into adipose tissue [19 - 22]. Due to great nutritional importance, the daily intake of linoleic acid at 17g and 12g for adult men and women has been recommended by the

Institute of Medicine of the National Academies [1]. This important issue has reinforced the awareness of food manufacturers and nutritionists to control FA profile in foodstuffs.

POLYUNSATURATED FATTY ACIDS (PUFAS)

PUFAs have been widely studied for their role in human health as well as for growth and development. PUFAs play a significant role in the prevention and treatment of cancer, CHD and inflammatory diseases [23 - 25]. Fish and some vegetable oils such as flaxseed, canola and soybean are the rich sources of PUFAs. These lower the overall cholesterol level as well as HDL and cardiovascular risk [26 - 28].

PUFAs especially linoleic and linolenic acid are considered to be essential FA since the human body is incapable to synthesize them, although these can be metabolized to longer-chain derivatives [29 - 31]. Linoleic and linolenic acids serve as substrates for the production of longer chain PUFAs which are constituents of cellular structure and as precursors for the production of glycerolipids and eicosanoids.

It has been also observed that shortage of α-linolenic acid is linked with poor growth, neurological abnormalities and scaly dermatitis [1]. Most important type of PUFA is omega-3 FAs, which help to decrease the threat of heart disease, neurogenerative disorders, control blood pressure, protector alongside plaque formed in blood vessel, and also play a role in brain development [32, 33].

Effects of diets rich in saturated and PUFAs on metabolic pathways examined and their proper ratios were suggested as a remedial measure in defensive medicine to lower serum cholesterol [34]. Daily intake of PUFA should be part of a diet, but maximum doze may be harmful to the nervous and digestion systems [35]. Therefore, recommended daily intake values ofα-linolenic acid for adult men and for women are 1.6g and 1.1g, respectively [36].

TRANS FATTY ACIDS (TFAS)

According to FDA, unsaturated fatty acids (UFAs) which contain one or more double bonds in a *trans* configuration are called TFAs. Regulatory authorities in some countries passed legislation to reduce the content of *trans* fat in their food supplies as minimum as possible by mandatory *trans* fat labeling on packed foods. Various commercial foods are high in TFAs [38], as they are produce during hydrogenation process, when liquid vegetable oil is converted into solid [37 - 39]. The hydrogenated fats become rancid very slowly, and that is much advantageous to the marketable food industries in making foods with a long shelf life.

Generally, partial hydrogenation improves the thermal stability to reduce oxidative process in contrast to unsaturated oils. But the natural kink found in cis-fatty acids vanishes and the molecule becomes linear and thus its physical properties match with SFAs. TFAs produce multiple adverse effects on CVD, inflammation, endothelial health, insulin sensitivity, oxidative stress, blood lipids, body weight and cancer [40, 41]. According to American Heart Association [42], the reason of death in the America is three forms of CVD, accounting about 38% of all deaths in 2002, the mainly predominant is CHD (heart attack), strokes and atherosclerosis. Epidemiologic studies revealed a connection among intake of TFAs and high risk of CHD [40, 43 - 45]. It was also observed that greater level of TFAs in the diet resulted in enhanced small dense LDL particles as compared to a diet rich in SFAs [40]. It was predicated that 2.5g of *trans* fat would be associated with 15% increase in CHD risk [46, 47]. In addition to that, association of C18:2*t* with the higher levels of plasma phospholipids were found to be potential risk of deadly ischemic heart disease (IHD) [48, 49]. Semi-quantitative food-frequency questionnaires revealed that tumor necrosis factor receptor (TNFR) levels has positive association with TFA in healthy women [40], plasma markers of systemic inflammation and endothelial dysfunction such as C-reactive protein, interlukin-6, E-selectin, tumor necrosis factor receptors and vascular cell adhesion molecules [50]. In some studies, it has been reported that dietary TFAs also increase the risk of type II diabetes [51, 52].

According to several studies, industrial TFAs have bad health effects, predominantly on blood lipoprotein profiles, diabetes, CHD and cancer whereas no information is presented on any positive health impact [53 - 56]. TFA eradication policies imposed enormous potential to reduce socioeconomic inequalities from CHD and decrease in mortality from NCDs [57].

New technologies are emerging to replace PHFs in foods to control the original structure of FAs and formation of TFAs in foods [50]. Laaksonen *et al.*, suggested that the fat quality is more imperative than the quantity in reducing risk of CVD [58]. Any product which contains TFA higher than or equivalent to 0.5 g per serving, it must be labeled on a separate line on the Nutritional Facts Panel whereas less than 0.5g per serving can be expressed as 0 g TFA. Therefore, many professional committees have been extensively engaged to make the certify-based statements which restrict dietary TFAs intake [59 - 62].

ROLE OF ANALYTICAL TECHNIQUES IN FATTY ACID COMPOSITION (FAC)

The appropriate analytical method for *trans* fat labeling on food products proposed by FDA listed in the Official Methods of Analysis of the AOAC

International (AOAC 2005) [63] or the Official Methods and Recommended Practices of the American Oil Chemists' Society (AOCS 2009) [64]. FDA regulations do not specify a particular method for the *trans* fat determination or other FAs in foods or edible oils [65]. Several methods have been used for the quantification of FAC together with TFAs in oils and fats as well as foodstuffs. The detailed analysis of lipids for fatty acid profile requires accurate, reliable and reproducible methods. Analytical techniques mostly utilized for the determination of FAC including TFAs are gas chromatography (GC), nuclear magnetic resonance (NMR) and infrared (IR) spectroscopy.

GAS CHROMATOGRAPHY

The analysis of *trans* fats by GC involves the chromatographic separation, identification and quantification of individual TFAs. The most widely employed detector in GC technique for the quantitation of FAs and TFAs is flame ionization detector (FID) [66].

Generally, for the quantitation of FAs including TFAs, the most used procedures are lipid extraction followed by conversion of the FAs into their respective methyl esters. Preparation of fatty acid methyl esters (FAMEs) from lipids carried out either by basic hydrolysis or by transesterification with acid or base catalyzed reactions. But, each of these methods has their own merits and demerits [67].

Koning *et al.* [68], determined the cis/*trans* distribution and compared the performance of the method with already reported method in which BF_3 and sodium methylate were used. Better separation of FAMEs was achieved on a 100 m SP-2560 column [63]. Although method is suitable for the determination of saturated and unsaturated fat in foods, but is unable to separate individual TFAs. Therefore, many other GC columns were used for the separation of TFAs and different trends of separation were observed by the columns coated with diverse polarity of stationary phases [69, 70].

Similarly, methylation processes have also been improved by many scientists [71 - 75]. But consequently single methylation procedure was not found sufficient to get the desirable derivative due to the complex nature of food products. GC had become extensively accepted as a suitable tool in the field of lipids [76], but preparation of volatile derivatives is required so as to optimize separation and decrease analysis time, especially with highly polar long capillary columns. When appropriate column such as AT-Silar-90, CP-Sil88 or SP–2560 with optimized operating conditions, capillary column will give a good separation of isomers (cis and *trans*) [45, 77]. Due to the high cost and limited availability of the FA standards and the complexity of the FAC, it is very intricate to recognize some FA peaks with the conventional FID detector [45]. To overcome this problem chemometric approach using Partial Least Square (PLS) has been successfully applied to get accurate results of FAs from GC data [2]. Applications of comprehensive two-dimensional GC-GC for triglyceride analysis have also been reported, but the applicability of this technique is limited [78]. GC coupled with Mass Spectrometer (MS) detector has been reflected as the "gold standard" technique for the quantitation of FAs, especially distinguishes cis and *trans* geometrical isomers. Separation of preliminary peaks in this technique is carried out by GC followed by MS recognition based on molecular ion peaks [79 - 82]. An interesting study about mass spectroscopic features described the capability of EI using high electron energy (70 eV) to recognize the geometry of double bond in polyenes of FAMEs. A minor variation was observed in this study between the geometries of cis and *trans* isomers of the terminal double bonds of trienes at peak intensity m/z 108, 150, and 192 recognizing the ω-3, ω-6, and ω-9 positions, respectively [83]. In the literature identification of the α-linolenic geometrical isomers was also reported by low electron energy (30 eV) mass spectrometer. The geometry of cis and *trans* isomers was identified by base peaks in the mass spectrum at m/z 79 and m/z 95, respectively [84].

TRANS FAT IN FOOD SAMPLES

Fat is the important component of biscuits which is responsible to maintain the quality of formulation. It has many roles. For example, in combination with other components it develops mouthfeel, texture and sense of lubricity of the product. But, unfortunately the choice of a fat for biscuits manufacturing commonly depends on the basis of economical and technological parameters, without taking into consideration the nutritional implications and quality of fat in most of the industries. In Table **1**, amount of fat contents and *trans* fat determined by GC-MS in biscuits, chocolates, pastries, chips and margarines are provided [79 - 82]. The quantity of total fat contents in the biscuits ranged from 13.7 to 27.6%, these levels are comparable to the reported values of fat content of Turkish biscuits 8.5 to 26.0% [85]. The amount of *trans* fat content in the biscuits ranged from 9.3 to 34.9% which is higher than those reported in the other studies [85 - 88].

Total fat contents of the chocolate ranged from 15.48 to 29.52%. The amount of *trans* fat content in the chocolate ranged from 4.56 to 8.49. The amount of *trans* fat in Turkish chocolates ranged at 1.85-3.68% [89], whereas New Zealand chocolate ranged at 0-3.1% [90].

Total fat contents of the pastry ranged from 28.45 to 38.75%. According to reported value of fat contents of different types of Austrian pastries (puff and flaky) almost similar fat content reported 27.89 to 38.87% [91]. The amount of *trans* fat content in the pastry ranged from 3.92 to 10.17%. However, in comparison, the amount of *trans* fat in Austrian pastry ranged at 0.10 to 9.91% [91]. Similarly, total fat and *trans* contents of the chips samples ranged from 18.72 to 37.13% and 4.91 to 14.13%, respectively [81]. A typical GC-MS chromatogram of FAMEs of chips oil is shown in Fig. (**1**). Whereas, amount of *trans* fat content in margarine ranged from 2.2 to 34.8%. According to other reported studies, few margarine samples were free from *trans* fat or had less than 1% *trans* fat [92 - 94]. The occurrence of higher *trans* fat level in the food samples is clear indication that partially hydrogenated oil has been used in the preparation process.

Table 1. Fat contents and *trans* values of some food samples by GC-MS.

Samples		Fat content (%)	Trans fat (%)
Biscuit	B-S1	20.70	9.26
	B-S2	25.69	28.83
	B-S3	17.14	29.26
	B-S4	21.90	34.88
	B-S5	13.69	14.46
	B-S6	23.70	27.64
	B-S7	20.60	32.60
	B-S8	15.82	33.97
	B-S9	20.59	29.88
	B-S10	26.73	26.92
	B-S11	27.64	27.57
	B-S12	26.85	25.65
Chocolate	Choc-S1	15.48	5.40
	Choc-S2	19.45	5.02
	Choc-S3	25.14	8.49
	Choc-S4	23.54	6.60
	Choc-S5	29.52	4.56
	Choc-S6	21.42	5.40
Pastry	Pas-S1	36.52	10.17
	Pas-S2	30.15	8.61
	Pas-S3	28.45	7.33
	Pas-S4	38.75	5.08
	Pas-S5	29.48	3.92

(Table 1) contd.....

Samples		Fat content (%)	Trans fat (%)
Chips	Ch-S1	21.15	13.73
	Ch-S2	19.75	5.13
	Ch-S3	33.5	11.52
	Ch-S4	25.48	4.91
	Ch-S5	29.85	7.62
	Ch-S6	37.13	13.72
	Ch-S7	24.61	11.87
	Ch-S8	28.35	14.13
	Ch-S9	31.54	11.0
	Ch-S10	34.85	8.26
	Ch-S11	18.72	8.96
	Ch-S12	25.41	9.14
Margarine	Mar-S1	-	26.40
	Mar-S2	-	2.20
	Mar-S3	-	34.80
	Mar-S4	-	32.60
	Mar-S5	-	26.50
	Mar-S6	-	11.50
	Mar-S7	-	8.30
	Mar-S8	-	11.70
	Mar-S9	-	15.60
	Mar-S10	-	30.30

HPLC AND LC/MS

Cis and *trans* isomers of FAs were also separated by high performance liquid chromatography (HPLC) technique using silver-ion stationary phases. But, contamination of traces of silver salts and the reproducibility are disadvantages associated with HPLC [95]. Therefore, separation of FAMEs carried out using small cartridges filled with a bonded benzene sulfonate medium incorporated with silver ions. Good separation was achieved for saturated and unsaturated fatty acids as well as geometrical isomers [96, 97]. Some researchers applied reversed-phase (RP) HPLC for the separation of cis and *trans* isomers. But, multiple usages of columns and need of gradient elution with water and polar organic solvents are drawbacks associated with this technique [98, 99]. Generally, in HPLC retention time of FAs depends on the total number of carbon atoms and double bonds present in the structure of FAs [98, 99]. Benefit of using HPLC over GC analysis is that lipids could be separated directly without derivatization procedure.

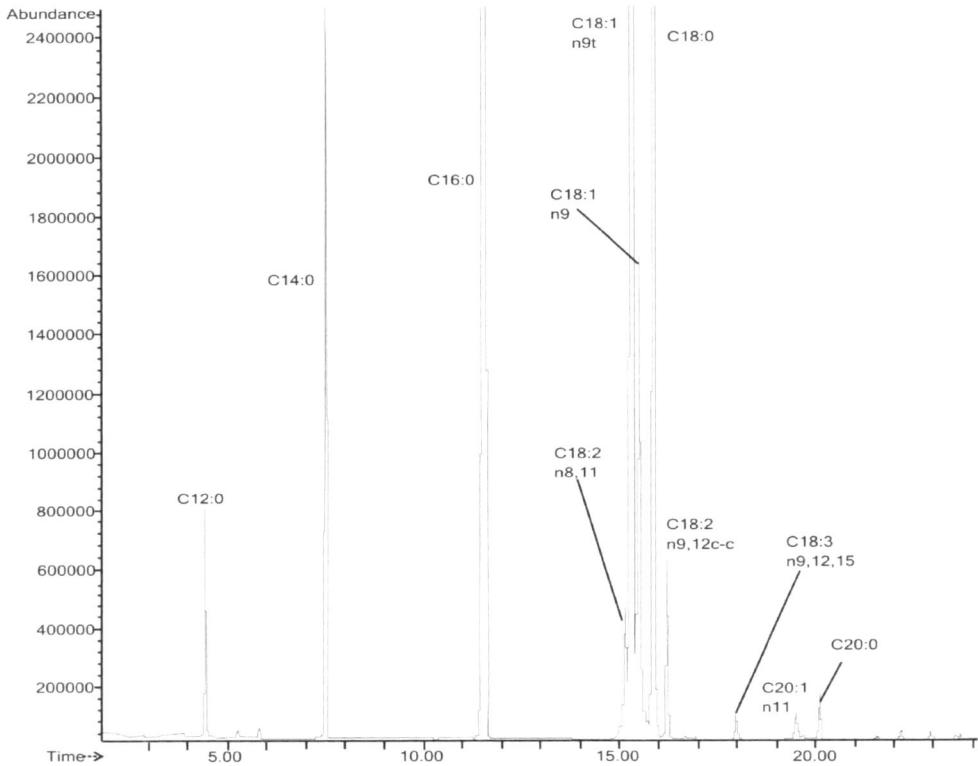

Fig. (1). Representative GC-MS Chromatogram of fatty acid methyl ester of chips oil.

RP-HPLC attached with MS is another analytical technique for the separation of cis and *trans* isomers of intact lipid and hydrolyzed FAs [98, 100 - 103]. RP-LC holds the great benefit to separate lipids on the basis of hydrophobicity of the alkyl group and polarity of the head group. But due to the use of large amount of solvents, cost of analysis and environmental issues, RP-HPLC/MS is not so frequently used for the quantitation of FAs and cis-*trans* ratio of lipids.

NUCLEAR MAGNETIC RESONANCE (NMR) SPECTROSCOPY

Often, GC is applied for the determination of FAC in edible fats and oils, but procedure is very laborious and lengthy. Furthermore, many toxic chemicals are involved in the procedures of saponification and esterification of the FAs. NMR technique has investigative prospective for distinguishing lipid classes. Both ¹H and ¹³C NMR have been explored for quantitative distinguishing between saturated, monounsaturated, and polyunsaturated components as well as cis and *trans* unsaturation in fats and oils. Restrictions of the NMR are associated to the sensitivity [104 - 106]. However,¹H NMR has advantage of sensitivity with respect to ¹³C NMR. Also high cost of NMR instrumentation is another drawback to its frequent application.

High resolution (HR) NMR spectroscopy has been emerged as a potential tool in the recent years to offer more information about the nature of constituents present in oils and fats than other spectroscopic techniques [107, 108]. For example, from a single NMR spectrum many parameters could be determined simultaneously without any sample pretreatment. Therefore, each group of fatty acid *i.e.*, saturated, monounsaturated, polyunsaturated and TFAs gives specific NMR signals in both ¹H and ¹³C NMR spectra, which consequently could be used for quantification of different classes of fats including *trans* fat [109 - 114]. NMR of vegetable oil samples offers more rapid evaluation of TFA, because this technique does not need any extraction, derivatization and sample preparation steps [115].

FOURIER TRANSFORM INFRARED (FT-IR) SPECTROSCOPY

In the past IR spectroscopy was considered as a qualitative technique for only determination of functional groups, but recently FT-IR spectroscopy is being used for the quantitative purposes as well with the support of chemometric tools. For the categorization of samples, the options include search for standards, distance match, similar match, discrimination analysis and QC compare search. TFAs in hydrogenated oils were determined by dispersive IR spectroscopy [116]. Although it is very simple method but did not provide detailed information about the nature of the *trans* isomers, such as the number of double bonds or pertinence the chain length. Later on Mossoba *et al.*, demonstrated that the pertinences of IR-ATR spectroscopy method was inadequate to >5% TFAs [117, 118].

Sedman and her group developed a method using heated single-bounce horizontal attenuated total reflectance (SB-HATR) accessory for the quantitation of *trans* fat content of neat fats and oils [119]. In another study Fourier transform near-infrared (FT-NIR) spectroscopy coupled with PLS calibration was applied to determine the *trans* fats of edible fats and oils by disposable glassvials (8-mm) for sample handling and measurements [120]. The accuracy of the predictions from calibration was not notably enhanced by supplement the base training set with samples and resulted in predictive errors while non-representative samples were studied. A rapid method was reported using FT-NIR for classification and quantification of the FAC of edible fats and oils. Spectra illustrated distinctive fingerprints for SFAs, cis and *trans* FAs, including long chain n-3 and n-6 PUFA. The quantitative models (chemometric approach) were based on adding exact data obtained from GC and FT-NIR spectral information into the calibration model [121]. Wu *et al.*, proposed a methodology for simultaneous determination of α-linolenic and linoleic acid in edible vegetable oils and their blends using NIR spectrometry. The obtained results indicated that developed method was practicable [122]. Moros *et al.*, determined the cis unsaturation and TFAs together with FFAs using ATR-FT-IR technique. A critical evaluation of edible oils between different multivariate calibrations was built based on the selected wave numbers or full spectra using two PLS algorithms [123].

Cereal-based Foods

The total fat contents and *trans* fat determined by FT-IR in different cereal-based foods are given in Table **2** [124]. Representative FT-IR spectrum of extracted oil from cereal-based food and expanded spectra of *trans* band between 990-940 cm⁻¹ of prepared trielaidine standards in canola oil is shown in Fig. (**2**).

Table 2. Oil contents and *trans* values of cereal based food samples by FT-IR.

Samples		Oil content (%)	Trans fat (%)
Cereal based foods	C-S1	12.7	13.5
	C-S2	23.9	8.9
	C-S3	17.5	11.6
	C-S4	15.8	16.3
	C-S5	21.3	3.3
	C-S6	27.2	13.4
	C-S7	14.8	10.4
	C-S8	8.2	2.5
	C-S9	20.1	3.9
	C-S10	23.6	15.0
	C-S11	16.2	9.1
	C-S12	19.2	13.0
	C-S13	10.3	6.7
	C-S14	26.1	16.1
	C-S15	13.4	10.0

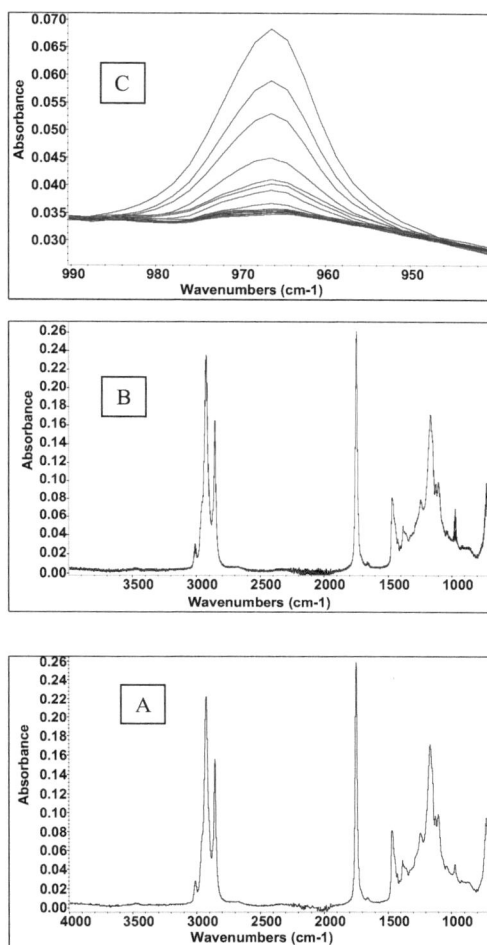

Fig. (2). (A) Representative FT-IR spectrum of neat lipid extract of cereal-based food **(B)** Group spectra of standards (canola oil spiked with trielaidine, 0.02–17.3%) **(C)** Expanded group spectrum of *trans* band absorption (from 990 to 940 cm^{-1}) of prepared standards.

In analyzed cereal-based food samples, the oil content was found to be in the range between 8.2–27.2%. Most of the samples contained more than 10% *trans* fat. In comparison to other available data on *trans* fat in cereal-based foods, the amount of *trans* fat reported in this study was much higher than breakfast cereals of Austria (0.2%) [91], with the exception of few samples, rest were within the range or even lower than the cereal-based products sold in America *i.e.* 7.2–10.7% and 3.1–15.5%, reported in two different studies [125, 126], while rest of the samples were higher. Daglioglu *et al.,* reported that Turkish cereal-based foods contain *trans* fats in the range of 0.1–31.0% [127], whereas Tavella *et al.* have reported *trans* fat in Argentinean cereal foods in the range of 7.62–11.07% [128].

CONSENT FOR PUBLICATION

Not applicable.

CONFLICT OF INTEREST

The author (editor) declares no conflict of interest, financial or otherwise.

ACKNOWLEDGMENT

Authors would like to thank National Centre of Excellence in Analytical Chemistry, University of Sindh, Jamshoro, Pakistan for providing research facilities and support to complete the present book chapter.

REFERENCES

[1]   Dennis S, Maury B, Tom D, Bob D, Jeffrey F, Brent F, *et al.* Food fats and oils. 9th ed. 2006; pp. 1-44.http://www.iseo.org/foodfats.pdf

[2]   Hajimahmoodi M, Vander Heyden Y, Sadeghi N, Jannat B, Oveisi MR, Shahbazian S. Gas-chromatographic fatty-acid fingerprints and partial least squares modeling as a basis for the simultaneous determination of edible oil mixtures. Talanta 2005; 66(5): 1108-16.
[http://dx.doi.org/10.1016/j.talanta.2005.01.011] [PMID: 18970097]

[3]   Bendsen NT, Chabanova E, Thomsen HS, *et al.* Effect of *trans* fatty acid intake on abdominal and liver fat deposition and blood lipids: a randomized trial in overweight postmenopausal women. Nutr Diabetes 2011; 1: e4.
[http://dx.doi.org/10.1038/nutd.2010.4] [PMID: 23154296]

[4]   Scarborough P, Rayner M, van Dis I, Norum K. Meta-analysis of effect of saturated fat intake on cardiovascular disease: overadjustment obscures true associations. Am J Clin Nutr 2010; 92(2): 458-9.
[http://dx.doi.org/10.3945/ajcn.2010.29504] [PMID: 20534750]

[5]   Esrey KL, Joseph L, Grover SA. Relationship between dietary intake and coronary heart disease mortality: lipid research clinics prevalence follow-up study. J Clin Epidemiol 1996; 49(2): 211-6.
[http://dx.doi.org/10.1016/0895-4356(95)00066-6] [PMID: 8606322]

[6]   Mann JI, Appleby PN, Key TJ, Thorogood M. Dietary determinants of ischaemic heart disease in health conscious individuals. Heart 1997; 78(5): 450-5.
[http://dx.doi.org/10.1136/hrt.78.5.450] [PMID: 9415002]

[7]   Erkkilä A, de Mello VDF, Risérus U, Laaksonen DE. Dietary fatty acids and cardiovascular disease: an epidemiological approach. Prog Lipid Res 2008; 47(3): 172-87.
[http://dx.doi.org/10.1016/j.plipres.2008.01.004] [PMID: 18328267]

[8]   Kris-Etherton PM, Yu S. Individual fatty acid effects on plasma lipids and lipoproteins: human studies. Am J Clin Nutr 1997; 65(5) (Suppl.): 1628S-44S.
[http://dx.doi.org/10.1093/ajcn/65.5.1628S] [PMID: 9129503]

[9]   Rivellese AA, Maffettone A, Vessby B, *et al.* Effects of dietary saturated, monounsaturated and n-3 fatty acids on fasting lipoproteins, LDL size and post-prandial lipid metabolism in healthy subjects. Atherosclerosis 2003; 167(1): 149-58.
[http://dx.doi.org/10.1016/S0021-9150(02)00424-0] [PMID: 12618280]

[10]   Rogers MA. Novel structuring strategies for unsaturated fats - Meeting the zero *trans*, zero-saturated

fat challenge: A review. Food Res Int 2009; 42: 747-53.
[http://dx.doi.org/10.1016/j.foodres.2009.02.024]

[11]   Mensink RP, Zock PL, Kester AD, Katan MB. Effects of dietary fatty acids and carbohydrates on the ratio of serum total to HDL cholesterol and on serum lipids and apolipoproteins: a meta-analysis of 60 controlled trials. Am J Clin Nutr 2003; 77(5): 1146-55.
[http://dx.doi.org/10.1093/ajcn/77.5.1146] [PMID: 12716665]

[12]   Papandreou D, Rousso I, Malindretos P, *et al.* Are saturated fatty acids and insulin resistance associated with fatty liver in obese children? Clin Nutr 2008; 27(2): 233-40.
[http://dx.doi.org/10.1016/j.clnu.2007.11.003] [PMID: 18234396]

[13]   Calder PC, Ahluwalia N, Brouns F, *et al.* Dietary factors and low-grade inflammation in relation to overweight and obesity. Br J Nutr 2011; 106 (Suppl. 3): S5-S78.
[http://dx.doi.org/10.1017/S0007114511005460] [PMID: 22133051]

[14]   Suganami T, Tanimoto-Koyama K, Nishida J, *et al.* Role of the Toll-like receptor 4/NF-kappaB pathway in saturated fatty acid-induced inflammatory changes in the interaction between adipocytes and macrophages. Arterioscler Thromb Vasc Biol 2007; 27(1): 84-91.
[http://dx.doi.org/10.1161/01.ATV.0000251608.09329.9a] [PMID: 17082484]

[15]   Schwingshackl L, Hoffmann G. Monounsaturated fatty acids and risk of cardiovascular disease: synopsis of the evidence available from systematic reviews and meta-analyses. Nutrients 2012; 4(12): 1989-2007.
[http://dx.doi.org/10.3390/nu4121989] [PMID: 23363996]

[16]   Kris-Etherton PM. AHA Science Advisory. Monounsaturated fatty acids and risk of cardiovascular disease. Circulation 1999; 100(11): 1253-8.
[http://dx.doi.org/10.1161/01.CIR.100.11.1253] [PMID: 10484550]

[17]   Harris WS. n-3 fatty acids and serum lipoproteins: human studies. Am J Clin Nutr 1997; 65(5) (Suppl.): 1645S-54S.
[http://dx.doi.org/10.1093/ajcn/65.5.1645S] [PMID: 9129504]

[18]   Hunter JE. Studies on effects of dietary fatty acids as related to their position on triglycerides. Lipids 2001; 36(7): 655-68.
[http://dx.doi.org/10.1007/s11745-001-0770-0] [PMID: 11521963]

[19]   Matulka RA, Noguchi O, Nosaka N. Safety evaluation of a medium- and long-chain triacylglycerol oil produced from medium-chain triacylglycerols and edible vegetable oil. Food Chem Toxicol 2006; 44(9): 1530-8.
[http://dx.doi.org/10.1016/j.fct.2006.04.004] [PMID: 16753249]

[20]   Nosaka N, Maki H, Suzuki Y, *et al.* Effects of margarine containing medium-chain triacylglycerols on body fat reduction in humans. J Atheroscler Thromb 2003; 10(5): 290-8.
[http://dx.doi.org/10.5551/jat.10.290] [PMID: 14718746]

[21]   Wall R, Ross RP, Fitzgerald GF, Stanton C. Fatty acids from fish: the anti-inflammatory potential of long-chain omega-3 fatty acids. Nutr Rev 2010; 68(5): 280-9.
[http://dx.doi.org/10.1111/j.1753-4887.2010.00287.x] [PMID: 20500789]

[22]   Simopoulos AP. Evolutionary aspects of diet: the omega-6/omega-3 ratio and the brain. Mol Neurobiol 2011; 44(2): 203-15.
[http://dx.doi.org/10.1007/s12035-010-8162-0] [PMID: 21279554]

[23]   Simopoulos AP. Omega-3 fatty acids in health and disease and in growth and development. Am J Clin Nutr 1991; 54(3): 438-63.
[http://dx.doi.org/10.1093/ajcn/54.3.438] [PMID: 1908631]

[24]   Drevon CA. Marine oils and their effects. Nutr Rev 1992; 50(4 (Pt 2)): 38-45.
[PMID: 1608564]

[25]   Horrocks LA, Yeo YK. Health benefits of docosahexaenoic acid (DHA). Pharmacol Res 1999; 40(3):

211-25.
[http://dx.doi.org/10.1006/phrs.1999.0495] [PMID: 10479465]

[26]   Kris-Etherton PM, Hecker KD, Binkoski AE. Polyunsaturated fatty acids and cardiovascular health. Nutr Rev 2004; 62(11): 414-26.
[http://dx.doi.org/10.1111/j.1753-4887.2004.tb00013.x] [PMID: 15622714]

[27]   Bonafini S, Antoniazzi F, Maffeis C, Minuz P, Fava C. Beneficial effects of ω-3 PUFA in children on cardiovascular risk factors during childhood and adolescence. Prostaglandins Other Lipid Mediat 2015; 120: 72-9.
[http://dx.doi.org/10.1016/j.prostaglandins.2015.03.006] [PMID: 25834924]

[28]   Hunter KA, Crosbie LC, Weir A, Miller GJ, Dutta-Roy AK. A residential study comparing the effects of diets rich in stearic acid, oleic acid, and linoleic acid on fasting blood lipids, hemostatic variables and platelets in young healthy men. J Nutr Biochem 2000; 11(7-8): 408-16.
[http://dx.doi.org/10.1016/S0955-2863(00)00097-8] [PMID: 11044636]

[29]   Judd JT, Baer DJ, Clevidence BA, *et al.* Effects of margarine compared with those of butter on blood lipid profiles related to cardiovascular disease risk factors in normolipemic adults fed controlled diets. Am J Clin Nutr 1998; 68(4): 768-77.
[http://dx.doi.org/10.1093/ajcn/68.4.768] [PMID: 9771853]

[30]   Frishman WH. Biologic markers as predictors of cardiovascular disease. Am J Med 1998; 104(6A): 18S-27S.
[http://dx.doi.org/10.1016/S0002-9343(98)00184-3] [PMID: 9684848]

[31]   Hilpert KF, West SG, Kris-Etherton PM, Hecker KD, Simpson NM, Alaupovic P. Postprandial effect of n-3 polyunsaturated fatty acids on apolipoprotein B-containing lipoproteins and vascular reactivity in type 2 diabetes. Am J Clin Nutr 2007; 85(2): 369-76.
[http://dx.doi.org/10.1093/ajcn/85.2.369] [PMID: 17284731]

[32]   Virginie D, Sylvie B, Michel L, Jacques F, Michel P. Fatty acid profiles of 80 vegetable oils with regard to their nutritional potential. Eur J Lipid Sci Technol 2007; 109: 710-32.
[http://dx.doi.org/10.1002/ejlt.200700040]

[33]   Youdim KA, Martin A, Joseph JA. Essential fatty acids and the brain: possible health implications. Int J Dev Neurosci 2000; 18(4-5): 383-99.
[http://dx.doi.org/10.1016/S0736-5748(00)00013-7] [PMID: 10817922]

[34]   Diniz YS, Cicogna AC, Padovani CR, *et al.* Diets rich in saturated and polyunsaturated fatty acids: metabolic shifting and cardiac health. Nutrition 2004; 20(2): 230-4.
[http://dx.doi.org/10.1016/j.nut.2003.10.012] [PMID: 14962692]

[35]   Wijendran V, Hayes KC. Dietary n-6 and n-3 fatty acid balance and cardiovascular health. Annu Rev Nutr 2004; 24: 597-615.
[http://dx.doi.org/10.1146/annurev.nutr.24.012003.132106] [PMID: 15189133]

[36]   Dietary Reference Intakes-Energy DRI. Carbohydrate, Fiber, Fat, Fatty Acids, Cholesterol, Protein, and Amino Acids. Washington, DC: The National Academies Press 2000.

[37]   Laurent L. Laure du C, Landy R, Lionel L. Fatty acid content of foods and intake levels in France. Eur J Lipid Sci Technol 2007; 109: 918-29.
[http://dx.doi.org/10.1002/ejlt.200600278]

[38]   Salas-Salvadó J, Márquez-Sandoval F, Bulló M. Conjugated linoleic acid intake in humans: a systematic review focusing on its effect on body composition, glucose, and lipid metabolism. Crit Rev Food Sci Nutr 2006; 46(6): 479-88.
[http://dx.doi.org/10.1080/10408390600723953] [PMID: 16864141]

[39]   Dashti N, Feng Q, Franklin FA. Long-term effects of *cis* and *trans* monounsaturated (18:1) and saturated (16:0) fatty acids on the synthesis and secretion of apolipoprotein A-I- and apolipoprotein B-containing lipoproteins in HepG2 cells. J Lipid Res 2000; 41(12): 1980-90.

[PMID: 11108731]

[40]   Gebauer SK, Psota TL, Kris-Etherton PM. The diversity of health effects of individual *trans* fatty acid isomers. Lipids 2007; 42(9): 787-99.
[http://dx.doi.org/10.1007/s11745-007-3095-8] [PMID: 17694343]

[41]   Soares-Miranda L, Stein PK, Imamura F, *et al.* Trans-fatty acid consumption and heart rate variability in 2 separate cohorts of older and younger adults. Circ Arrhythm Electrophysiol 2012; 5(4): 728-38.
[http://dx.doi.org/10.1161/CIRCEP.111.966259] [PMID: 22772898]

[42]   American Heart Association. Heart Disease and Stroke Statistics. Dallas, Texas: Update 2005.

[43]   Noone EJ, Roche HM, Nugent AP, Gibney MJ. The effect of dietary supplementation using isomeric blends of conjugated linoleic acid on lipid metabolism in healthy human subjects. Br J Nutr 2002; 88(3): 243-51.
[http://dx.doi.org/10.1079/BJN2002615] [PMID: 12207834]

[44]   Flock MR, Kris-Etherton PM. Dietary Guidelines for Americans 2010: implications for cardiovascular disease. Curr Atheroscler Rep 2011; 13(6): 499-507.
[http://dx.doi.org/10.1007/s11883-011-0205-0] [PMID: 21874316]

[45]   Huang Z, Wang B, Crenshaw AA. A simple method for the analysis of *trans* fatty acid with GC-MS and AT(TM)-Silar-90 capillary column. Food Chem 2006; 98: 593-8.
[http://dx.doi.org/10.1016/j.foodchem.2005.05.013]

[46]   Lemaitre RN, King IB, Raghunathan TE, *et al.* Cell membrane *trans*-fatty acids and the risk of primary cardiac arrest. Circulation 2002; 105(6): 697-701.
[http://dx.doi.org/10.1161/hc0602.103583] [PMID: 11839624]

[47]   Risérus U, Berglund L, Vessby B. Conjugated linoleic acid (CLA) reduced abdominal adipose tissue in obese middle-aged men with signs of the metabolic syndrome: a randomised controlled trial. Int J Obes Relat Metab Disord 2001; 25(8): 1129-35.
[http://dx.doi.org/10.1038/sj.ijo.0801659] [PMID: 11477497]

[48]   Lemaitre RN, King IB, Mozaffarian D, *et al.* Plasma phospholipid *trans* fatty acids, fatal ischemic heart disease, and sudden cardiac death in older adults: the cardiovascular health study. Circulation 2006; 114(3): 209-15.
[http://dx.doi.org/10.1161/CIRCULATIONAHA.106.620336] [PMID: 16818809]

[49]   Jakobsen MU, Bysted A, Andersen NL, *et al.* Intake of ruminant *trans* fatty acids and risk of coronary heart disease-an overview. Atheroscler Suppl 2006; 7(2): 9-11.
[http://dx.doi.org/10.1016/j.atherosclerosissup.2006.04.004] [PMID: 16713389]

[50]   Tarrago-Trani MT, Phillips KM, Lemar LE, Holden JM. New and existing oils and fats used in products with reduced *trans*-fatty acid content. J Am Diet Assoc 2006; 106(6): 867-80.
[http://dx.doi.org/10.1016/j.jada.2006.03.010] [PMID: 16720128]

[51]   Salmerón J, Hu FB, Manson JE, *et al.* Dietary fat intake and risk of type 2 diabetes in women. Am J Clin Nutr 2001; 73(6): 1019-26.
[http://dx.doi.org/10.1093/ajcn/73.6.1019] [PMID: 11382654]

[52]   van Dam RM, Willett WC, Rimm EB, Stampfer MJ, Hu FB. Dietary fat and meat intake in relation to risk of type 2 diabetes in men. Diabetes Care 2002; 25(3): 417-24.
[http://dx.doi.org/10.2337/diacare.25.3.417] [PMID: 11874924]

[53]   Booker CS, Mann JI. *Trans* fatty acids and cardiovascular health: translation of the evidence base. Nutr Metab Cardiovasc Dis 2008; 18(6): 448-56.
[http://dx.doi.org/10.1016/j.numecd.2008.02.005] [PMID: 18468872]

[54]   Teegala SM, Willett WC, Mozaffarian D. Consumption and health effects of *trans* fatty acids: a review. J AOAC Int 2009; 92(5): 1250-7.
[PMID: 19916363]

[55]   Bendsen NT, Christensen R, Bartels EM, Astrup A. Consumption of industrial and ruminant *trans* fatty acids and risk of coronary heart disease: a systematic review and meta-analysis of cohort studies. Eur J Clin Nutr 2011; 65(7): 773-83.
[http://dx.doi.org/10.1038/ejcn.2011.34] [PMID: 21427742]

[56]   Brouwer IA, Wanders AJ, Katan MB. Effect of animal and industrial *trans* fatty acids on HDL and LDL cholesterol levels in humans--a quantitative review. PLoS One 2010; 5(3): e9434.
[http://dx.doi.org/10.1371/journal.pone.0009434] [PMID: 20209147]

[57]   Allen K, Pearson-Stuttard J, Hooton W, Diggle P, Capewell S, O'Flaherty M. Potential of *trans* fats policies to reduce socioeconomic inequalities in mortality from coronary heart disease in England: cost effectiveness modelling study. BMJ 2015; 351: h4583.
[http://dx.doi.org/10.1136/bmj.h4583] [PMID: 26374614]

[58]   Laaksonen DE, Nyyssönen K, Niskanen L, Rissanen TH, Salonen JT. Prediction of cardiovascular mortality in middle-aged men by dietary and serum linoleic and polyunsaturated fatty acids. Arch Intern Med 2005; 165(2): 193-9.
[http://dx.doi.org/10.1001/archinte.165.2.193] [PMID: 15668366]

[59]   WHO/FAO Expert Consultation. Diet, nutrition, and the prevention of chronic diseases 2003.

[60]   EURODIET. Nutrition & diet for healthy lifestyles in Europe. Science & Policy Implications 2000; pp. 1-17.

[61]   Netherlands HCOT. Guidelines for a healthy diet Publication no 2006/21, in The Hague: Health Council of the Netherlands 2006

[62]   American Diabetes Association. Nutrition recommendations and interventions for diabetes. Diabetes Care 2007; 30 (Suppl. 1): S48-65.
[http://dx.doi.org/10.2337/dc07-S048] [PMID: 17192379]

[63]   AOAC. Official methods of analysis. 18th ed., Gaithersburg, MD: AOAC International 2005.

[64]   Firestone D. Official methods and recommended practices of the AOCS. 6th ed., Urbana, IL: AOCS Press 2009.

[65]   Department of Health and Human Services. FDA. Food labeling: *trans* fatty acids in nutrition labeling: nutrient content claims, and health claims; final rule. Fed Regist 2003; 68(133): 41434-506.

[66]   Delmonte P, Rader JI. Evaluation of gas chromatographic methods for the determination of *trans* fat. Anal Bioanal Chem 2007; 389(1): 77-85.
[http://dx.doi.org/10.1007/s00216-007-1392-y] [PMID: 17572885]

[67]   Antolín EM, Delange DM, Canavaciolo VG. Evaluation of five methods for derivatization and GC determination of a mixture of very long chain fatty acids (C24:0-C36:0). J Pharm Biomed Anal 2008; 46(1): 194-9.
[http://dx.doi.org/10.1016/j.jpba.2007.09.015] [PMID: 18031966]

[68]   de Koning S, van der Meer B, Alkema G, Janssen H-G, Brinkman UAT. Automated determination of fatty acid methyl ester and cis/*trans* methyl ester composition of fats and oils. J Chromatogr A 2001; 922(1-2): 391-7.
[http://dx.doi.org/10.1016/S0021-9673(01)00926-8] [PMID: 11486889]

[69]   Ferreri C, Faraone Mennella MR, Formisano C, Landi L, Chatgilialoglu C. Arachidonate geometrical isomers generated by thiyl radicals: the relationship with *trans* lipids detected in biological samples. Free Radic Biol Med 2002; 33(11): 1516-26.
[http://dx.doi.org/10.1016/S0891-5849(02)01083-3] [PMID: 12446209]

[70]   Zghibeh CM, Raj Gopal V, Poff CD, Falck JR, Balazy M. Determination of *trans*-arachidonic acid isomers in human blood plasma. Anal Biochem 2004; 332(1): 137-44.
[http://dx.doi.org/10.1016/j.ab.2004.04.030] [PMID: 15301958]

[71]   Christie WW. Preparation of ester derivatives of fatty acids for chromatographic analysis. Eds Adv

Lipid Meth. Dundee, UK: The Oily Press 1993; pp. 69-110.

[72] Yurawecz MP, Hood JK, Roach JAG, Mossoba MM, Daniels DH, Ku Y, *et al.* Conversion of allylichydroxyoleate to conjugated linoleic acid and methoxyleate by acid-catalyzed methylation procedures. J Am Oil Chem Soc 1994; 71: 1149-55.
[http://dx.doi.org/10.1007/BF02675911]

[73] Kramer JKG, Fellner V, Dugan MER, Sauer FD, Mossoba MM, Yurawecz MP. Evaluating acid and base catalysts in the methylation of milk and rumen fatty acids with special emphasis on conjugated dienes and total *trans* fatty acids. Lipids 1997; 32(11): 1219-28.
[http://dx.doi.org/10.1007/s11745-997-0156-3] [PMID: 9397408]

[74] Yamasaki M, Kishihara K, Ikeda I, Sugano M, Yamada K. A recommended esterification method for gas chromatographic measurement of conjugated linoleic acid. J Am Oil Chem Soc 1999; 76: 933-8.
[http://dx.doi.org/10.1007/s11746-999-0109-0]

[75] Park Y, Albright KJ, Cai ZY, Pariza MW. Comparison of methylation procedures for conjugated linoleic acid and artifact formation by commercial (trimethylsilyl) diazomethane. J Agric Food Chem 2001; 49(3): 1158-64.
[http://dx.doi.org/10.1021/jf001209z] [PMID: 11312828]

[76] Jalali-Heravi M, Vosough M. Characterization and determination of fatty acids in fish oil using gas chromatography-mass spectrometry coupled with chemometric resolution techniques. J Chromatogr A 2004; 1024(1-2): 165-76.
[http://dx.doi.org/10.1016/j.chroma.2003.10.032] [PMID: 14753719]

[77] Ratnayake WMN, Plouffe LJ, Pasquier E, Gagnon C. Temperature-sensitive resolution of cis- and *trans*-fatty acid isomers of partially hydrogenated vegetable oils on SP-2560 and CP-Sil 88 capillary columns. J AOAC Int 2002; 85(5): 1112-8.
[PMID: 12374411]

[78] Haglund P, Harju M, Danielsson C, Marriott P. Effects of temperature and flow regulated carbon dioxide cooling in longitudinally modulated cryogenic systems for comprehensive two-dimensional gas chromatography. J Chromatogr A 2002; 962(1-2): 127-34.
[http://dx.doi.org/10.1016/S0021-9673(02)00433-8] [PMID: 12198957]

[79] Kandhro A, Sherazi STH, Mahesar SA, Bhanger MI, Younis Talpur M, Rauf A. GC-MS quantification of fatty acid profile including *trans* FA in the locally manufactured margarines of Pakistan. Food Chem 2008; 109(1): 207-11.
[http://dx.doi.org/10.1016/j.foodchem.2007.12.029] [PMID: 26054282]

[80] Kandhro A, Sherazi STH, Mahesar SA, Bhanger MI, Talpur MY, Arain S. Monitoring of fat content, free fatty acid and fatty acid profile including *trans* fat in Pakistani biscuits. J Am Oil Chem Soc 2008; 85: 1057-61.
[http://dx.doi.org/10.1007/s11746-008-1297-8]

[81] Kandhro A, Sherazi STH, Mahesar SA, Bhanger MI, Talpur MY, Latif Y. Variations of fatty acid composition including *trans* fat in commonly used different brands of potato chips by GC-MS. Pak J Anal Environ Chem 2010; 11: 36-41.

[82] Kandhro A, Sherazi STH, Rubina S, Razia S. Ambrat, Arfa Y. Consequence of fatty acids profile including *trans* fat in chocolate and pastry samples. Int Food Res J 2013; 20: 1337-41.

[83] Mjos SA, Pettersen J. Determination of *trans* double bonds in polyunsaturated fatty acid methyl esters from their electron impact mass spectra. Eur J Lipid Sci Technol 2003; 105: 156-64.
[http://dx.doi.org/10.1002/ejlt.200390031]

[84] Hejazi L, Ebrahimi D, Hibbert DB, Guilhaus M. Compatibility of electron ionization and soft ionization methods in gas chromatography/orthogonal time-of-flight mass spectrometry. Rapid Commun Mass Spectrom 2009; 23(14): 2181-9.
[http://dx.doi.org/10.1002/rcm.4131] [PMID: 19530152]

[85]   Daglioglu O, Murat T, Tuncel B. Determination of fatty acid composition and total *trans* fatty acids of Turkish biscuits by capillary gas-liquid chromatography. Eur Food Res Technol 2000; 211: 41-4.
[http://dx.doi.org/10.1007/s002170050586]

[86]   Caponio F, Summo C, Delcuratolo D, Pasqualone A. Quality of the lipid fraction of Italian biscuits. J Sci Food Agric 2006; 86: 356-61.
[http://dx.doi.org/10.1002/jsfa.2357]

[87]   Huang ZE, Wang B, Pace RD, Oh JH. fatty acid content of selected foods in an African–American community. J Food Sci 2006; 71: 322-7.
[http://dx.doi.org/10.1111/j.1750-3841.2006.00056.x]

[88]   Martin CA, Carapelli R, Visantainer JV, Matsushita M, de Souza NE. *Trans* fatty acid content of Brazilian biscuits. Food Chem 2005; 93: 445-8.
[http://dx.doi.org/10.1016/j.foodchem.2004.10.022]

[89]   Karabulut I. Fatty acid composition of frequently consumed foods in Turkey with special emphasis on *trans* fatty acids. Int J Food Sci Nutr 2007; 58(8): 619-28.
[http://dx.doi.org/10.1080/09637480701368967] [PMID: 17852509]

[90]   Saunders D, Jones S, Devane GJ, Scholes P, Lake RJ, Paulin SM. *Trans* fatty acids in the New Zealand food supply. J Food Compos Anal 2008; 21: 320-5.
[http://dx.doi.org/10.1016/j.jfca.2007.12.004]

[91]   Wagner KH, Plasser E, Proell C, Kanzler S. Comprehensive studies on the *trans* fatty acid content of Austrian foods: Convenience products, fast food and fats. Food Chem 2008; 108(3): 1054-60.
[http://dx.doi.org/10.1016/j.foodchem.2007.11.038] [PMID: 26065770]

[92]   Triantafillou D, Zografos V, Katsikas H. Fatty acid content of margarines in the Greek market (including *trans*-fatty acids): a contribution to improving consumers' information. Int J Food Sci Nutr 2003; 54(2): 135-41.
[http://dx.doi.org/10.1080/0963748031000084089] [PMID: 12701370]

[93]   Brat J, Pokorny J. Fatty acid composition of margarines and cooking fats available on the Czech market. J Food Compos Anal 2000; 13: 337-43.
[http://dx.doi.org/10.1006/jfca.1999.0877]

[94]   Karabulut I, Turan S. Some properties of margarines and shortenings marketed in Turkey. J Food Compos Anal 2006; 19: 55-8.
[http://dx.doi.org/10.1016/j.jfca.2004.06.016]

[95]   Tsuzuki W, Ushida K. Preparative separation of cis- and *trans*-isomers of unsaturated fatty acid methyl esters contained in edible oils by reversed-phase high-performance liquid chromatography. Lipids 2009; 44(4): 373-9.
[http://dx.doi.org/10.1007/s11745-008-3271-5] [PMID: 19089482]

[96]   Toschi TG, Capella P, Holt C, Christie WW. A comparison of silver ion HPLC plus GC with Fourier-trasnform IR spectroscopy for the determination of *trans* double bonds in unsaturated fatty acids. J Sci Food Agric 1993; 61: 261-6.
[http://dx.doi.org/10.1002/jsfa.2740610220]

[97]   Adlof RO, Copes LC, Emken EA. Analysis of the monoenoic fatty acid distribution in hydrogenated vegetable oils by silver-ion high-performance liquid chromatography. J Am Oil Chem Soc 1995; 72: 571-4.
[http://dx.doi.org/10.1007/BF02638858]

[98]   Juanéda P. Utilisation of reversed-phase high-performance liquid chromatography as an alternative to silver-ion chromatography for the separation of cis- and *trans*-C18:1 fatty acid isomers. J Chromatogr A 2002; 954(1-2): 285-9.
[http://dx.doi.org/10.1016/S0021-9673(02)00179-6] [PMID: 12058913]

[99]   Destaillats F, Golay PA, Joffre F, *et al.* Comparison of available analytical methods to measure *trans*-

octadecenoic acid isomeric profile and content by gas-liquid chromatography in milk fat. J Chromatogr A 2007; 1145(1-2): 222-8.
[http://dx.doi.org/10.1016/j.chroma.2007.01.062] [PMID: 17275831]

[100] Bird SS, Marur VR, Stavrovskaya IG, Kristal BS. Qualitative characterization of the rat liver mitochondrial lipidome using LC-MS profiling and high energy collisional dissociation (HCD) all ion fragmentation. Metabolomics 2013; 9(1) (Suppl.): 67-83.
[http://dx.doi.org/10.1007/s11306-012-0400-1] [PMID: 23646040]

[101] Bird SS, Marur VR, Stavrovskaya IG, Kristal BS. Separation of cis-*trans* phospholipid isomers using reversed phase LC with high resolution MS detection. Anal Chem 2012; 84(13): 5509-17.
[http://dx.doi.org/10.1021/ac300953j] [PMID: 22656324]

[102] Villegas C, Zhao Y, Curtis JM. Two methods for the separation of monounsaturated octadecenoic acid isomers. J Chromatogr A 2010; 1217(5): 775-84.
[http://dx.doi.org/10.1016/j.chroma.2009.12.011] [PMID: 20022011]

[103] Momchilova SM, Nikolova-Damyanova BM. Separation of isomeric octadecenoic fatty acids in partially hydrogenated vegetable oils as p-methoxyphenacyl esters using a single-column silver ion high-performance liquid chromatography (Ag-HPLC). Nat Protoc 2010; 5(3): 473-8.
[http://dx.doi.org/10.1038/nprot.2009.232] [PMID: 20203664]

[104] Sherazi STH. Qunatitative analysis of some vegetable oils by ^{13}C NMR and chromatographic methods. PhD Dissertation. University of Sindh, 1996.

[105] Shoolery JN. Some quntitative applications of ^{13}C NMR spectroscopy in progress in NMR spectroscopy. England: Pergamon Press 1977; 11: pp. 79-93.

[106] Shoolery JN. ^{13}C NMR as quantitaive analytical method for determining the composition of fats and oils. Application Note NMR-75-3.. Palo Alto, CA, USA: Varian Associates 1975.

[107] Hidalgo FJ, Zamora R. Edible oil analysis by high-resolution nuclear magnetic resonance spectroscopy: recent advances and future perspectives. Trends Food Sci Technol 2003; 14: 499-506.
[http://dx.doi.org/10.1016/j.tifs.2003.07.001]

[108] Mannina L, Sobolev AP, Segre AL. Olive oil as seen by NMR and chemometrics. Spectrosc Eur 2003; 15: 6-14.

[109] Miyake Y, Yokomizo K, Matsuzaki N. Determination of unsaturated fatty acid composition by high-resolution nuclear magnetic resonance spectroscopy. J Am Oil Chem Soc 1998; 75: 1091-4.
[http://dx.doi.org/10.1007/s11746-998-0118-4]

[110] Sacchi R, Addeo F, Paolillo L. ^{1}H and ^{13}C NMR of virgin olive oil. An overview. Magn Reson Chem 1997; 35: S133-45.
[http://dx.doi.org/10.1002/(SICI)1097-458X(199712)35:13<S133::AID-OMR213>3.0.CO;2-K]

[111] Guillén MD, Uriarte PS. Monitoring by ^{1}H nuclear magnetic resonance of the changes in the composition of virgin linseed oil heated at frying temperature. Comparison with the evolution of other edible oils. Food Control 2012; 28: 59-68.
[http://dx.doi.org/10.1016/j.foodcont.2012.04.024]

[112] Knothe G, Kenar JA. Determination of the fatty acid profile by ^{1}H-NMR spectroscopy. Eur J Lipid Sci Technol 2004; 106: 88-96.
[http://dx.doi.org/10.1002/ejlt.200300880]

[113] Ng S. Analysis of positional distribution of fatty acids in palm oil by ^{13}C NMR spectroscopy. Lipids 1985; 20: 778-82.
[http://dx.doi.org/10.1007/BF02534402]

[114] Wollenberg KF. Quantitative high resolution ^{13}C nuclear magnetic resonance of the olefinic and carbonyl carbons of edible vegetable oils. J Am Oil Chem Soc 1990; 67: 487-94.
[http://dx.doi.org/10.1007/BF02540753]

[115] Gunstone FD. The composition of hydrogenated fats by high-resolution ^{13}C nuclear magnetic resonance spectroscopy. J Am Oil Chem Soc 1993; 70: 965-70.
[http://dx.doi.org/10.1007/BF02543022]

[116] Firestone D, Sheppard A. Determination of *trans* fatty acids.Adv Lipid Meth-One. Ayr, UK: The Oily Press 1992; pp. 273-322.

[117] Mossoba MM, Adam M, Lee T, Bastyr J. Rapid determination of total *trans* fat content. An attenuated total reflection infrared spectroscopy international collaborative study. J AOAC Int 2001; 84(4): 1144-50.
[PMID: 11501916]

[118] Mossoba MM, Kramer JKG, Delmonte P, Yurawecz MP, Rader JI. Official Methods for the Determination of Trans Fat. Champaign, IL, USA: AOCS Press 2003.

[119] Sedman J, van de Voort FR, Ismail A. Simultaneous determination of iodine value and *trans* content of fats and oils by single-bounce horizontal attenuated total reflectance Fourier transform infrared spectroscopy. J Am Oil Chem Soc 2000; 77: 399-403.
[http://dx.doi.org/10.1007/s11746-000-0064-y]

[120] Li H, van de Voort FR, Ismail AA, Sedman J, Cox R. *Trans* determination of edible oils by Fourier transform near-infrared spectroscopy. J Am Oil Chem Soc 2000; 77: 1061-7.
[http://dx.doi.org/10.1007/s11746-000-0167-5]

[121] Azizian H, Kramer JK. A rapid method for the quantification of fatty acids in fats and oils with emphasis on *trans* fatty acids using Fourier transform near infrared spectroscopy (FT-NIR). Lipids 2005; 40(8): 855-67.
[http://dx.doi.org/10.1007/s11745-005-1448-3] [PMID: 16296405]

[122] Wu D, Chen X, Shi P, Wang S, Feng F, He Y. Determination of alpha-linolenic acid and linoleic acid in edible oils using near-infrared spectroscopy improved by wavelet transform and uninformative variable elimination. Anal Chim Acta 2009; 634(2): 166-71.
[http://dx.doi.org/10.1016/j.aca.2008.12.024] [PMID: 19185115]

[123] Moros J, Roth M, Garrigues S, Guardia MD. Preliminary studies about thermal degradation of edible oils through attenuated total reflectance mid-infrared spectrometry. Food Chem 2009; 114: 1529-36.
[http://dx.doi.org/10.1016/j.foodchem.2008.11.040]

[124] Mahesar SA, Kandhro AA, Cerretani L, Bendini A, Sherazi STH, Bhanger MI. Determination of total *trans* fat content in Pakistani cereal based foods by SB-HATR FT-IR spectroscopy coupled with partial least square regression. Food Chem 2010; 123: 1289-93.
[http://dx.doi.org/10.1016/j.foodchem.2010.05.101]

[125] Kim Y, Himmelsbach DS, Kays SE. ATR-Fourier transform mid-infrared spectroscopy for determination of *trans* fatty acids in ground cereal products without oil extraction. J Agric Food Chem 2007; 55(11): 4327-33.
[http://dx.doi.org/10.1021/jf063729l] [PMID: 17472389]

[126] Robinson JE, Singh R, Kays SE. Evaluation of an automated hydrolysis and extraction method for quantification of total fat, lipid classes and *trans* fat in cereal products. Food Chem 2008; 107: 1144-50.
[http://dx.doi.org/10.1016/j.foodchem.2007.09.041]

[127] Daglioglu O, Tasan M, Tuncel B. Determination of fatty acid composition and total *trans* fatty acids in cereal-based Turkish foods. Turk J Chem 2002; 26: 705-10.

[128] Tavella M, Peterson G, Espeche M, Cavallero E, Cipolla L, Perego L. *Trans* fatty acid content of a selection of foods in Argentina. Food Chem 2000; 69: 209-13.
[http://dx.doi.org/10.1016/S0308-8146(99)00257-5]

CHAPTER 10

Analysis of Adulterants in Saffron

Brijesh Sharma[*]

Institute of Biomedical and Education Research, Mangalayatan University, Beswan, Aligarh, 202146, India

Abstract: Saffron is of commercial importance for manufacturers because of its flavoring properties in food industry and medicinal properties in pharmaceutical industry. The biological source of saffron is the dried stigma of *Crocus sativus L.* flower. In order to decrease its cost and meet large demand across the globe; it is adulterated with other spurious materials by criminals to mislead the consumers which in turn renders it completely useless or even harmful. The genuine saffron should meet the quality requirements as laid down by FDA or ISO specifications. In the present chapter, adulteration of saffron and various physical, chemical and analytical techniques are described for the rapid discrimination between pure and impure saffron.

Keywords: FDA, ISO, Spectroscopy, Spurious, TLC.

INTRODUCTION

Saffron is the processed stigma of *Crocus sativus* flower Linnaeus. It is having tremendous value in spices as of its antiquity for color, flavor and medicinal properties [1]. The origin of saffron is still not clear but it is speculated that this plant first grew in the eastern Mediterranean, probably Asia Minor and Persia. The word"saffron" owes its origin to Arabic za´-faran means "be yellow." The Greeks called this spice by the name of Krokos but the name seems to be pre-Greek and may be from Babylonian- Assyrian origin [2]. An exact identification of Crocus sativus was reported in 1500 B.C., on the Greek island of Santorini [3]. From the eastern mediterranean region, the cultivation of saffron spread to other parts of world like Europe and Asia. Saffron **is** now produced by many countries like Spain, France, Italy, Germany, Iran and India [4]. In India, saffron producing states are Gujarat, Himachal Pradesh and Jammu and Kashmir [5]. The high price of saffron makes it a vulnerable and attractive object of adulteration among criminals and commercial producers.

[*] **Corresponding author Brijesh Sharma:** Institute of Biomedical and Education Research, Mangalayatan University, Beswan, Aligarh, 202146, India; Tel: 9897288856; E-mail: brijesh.sharma@mangalayatan.edu.in

Alankar Shrivastava (Ed.)

The forensic analysis of the saffron poses a challenge to the scientists for determination of authenticity and detection of adulterants. In this chapter, chemical constituents, therapeutic uses of saffron and analytical methods for identification of adulterants in it are discussed.

USES

Various industries like food, perfume, dye or ink industry had been using Saffron as spice owing to its flavoring and coloring properties since long back [6] and even now-a-days, some countries, *viz*. Italy, France and Spain are using this spice in traditional seafood dish. Saffron also finds its use in a few cakes, *e.g.,* the German saffron cake, popularly known as "Gugelhupf." Saffron has a wide spectrum of medicinal uses. A long time ago, Romans believed to get relief from its amazing ability to ward off hangovers by squeezing the spice in their wine [4]. Folk and ayurvedic systems of medicine quote Saffron as a sedative, expectorant, anti asthmatic, emmenagogue and adaptogenic agent besides its use in opioid preparations for pain relief from in16 -19th centuries [7] Recent pharmacological researches have confirmed its antitumor properties [8], free radical scavenging properties [9], hypolipaemic effects [10] and immense use in neurodegenerative diseases following memory impairment [11].

CHEMICAL COMPOSITION

A strong odor and natural dark yellow color make Saffron an important spicy herb. The characteristic color owes to the presence of degraded carotenoid substances, crocin [12] and crocetin [13] and the flavor is attributed to the carotenoid oxidation products, mainly safranal [14] while the bitterness originates from a glucoside picrocrocin. Phytochemical investigation has shown that saffron is composed of approximately moisture (10%), protein (12%), fat, minerals, crude fibre (5% each) and sugars (63%) including starch, reducing sugars, pentosans, gums, pectin, and dextrins (% w/w). Traces of vitamins like riboflavin and thiamine are also reported in saffron. [15] The amounts of all active principles (fig.1) [16] may vary greatly depending on the growing conditions and country of origin [17]. Table 1 shows the proximate analysis of Saffron.

Table 1. Proximate analysis of saffron [18].

S. No.	Component	Mass Percentage
1.	Water-soluble components	53.0
2.	Gums	10.0
3.	Pentosans	8.0
4.	Pectins	6.0

(Table 1) contd.....

S. No.	Component	Mass Percentage
5.	Starch	6.0
6.	Crocin	2.0
7.	Carotenoids	1.0
8.	Lipids	12.0
9.	Non-volatile oils	6.0
10.	Volatile oils	1.0
11.	Protein	12.0
12.	Inorganic matter ("ash")	6.0
13.	HCl soluble ash	0.5
14.	Water	10.0
15.	Fiber (Crude)	5.0

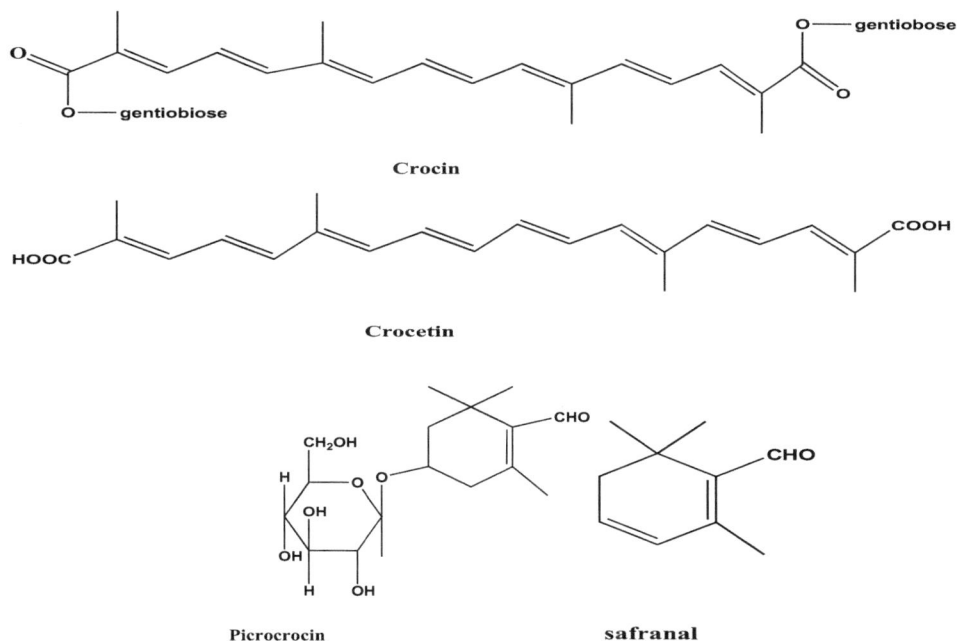

Crocin

Crocetin

Picrocrocin   safranal

Fig. (1). Chemical structure of active principles of Saffron.

QUALITY REQUIREMENTS

FDA rules [19] permit Saffron for dressing and flavoring of natural food in cooking processes, if it fulfills the following criteria:

i. Stigmas should be yellow in color and the foreign organic compounds must be less than 10%.
ii. The volatile compounds and humidity should not exceed 14% when the saffron is subjected to 100°C for drying.
iii. The maximum limit for total ash and soluble ash is 1% each.
iv. Microbial content of saffron usually consists of aerobic spore-producing bacteria like Bacillus, Fungi and rarely E. coli, Lactobacillus, Micrococcus, Staphylococcus, and Streptococcus.
v. Heating method of sterilization is not preferable since it affects the color, taste, and odor of the product. Gamma, microwave, and ultraviolet (UV) radiations in combination with fumigation with ethylene oxide seems be more helpful [20]. The likelihood of breakdown of pigments during gamma irradiation is taken care of [21].

ADULTERATION OF SAFFRON

The ever increasing price and demand frequently are the two main reasons for adulteration of Saffron. Adulteration of Saffron in the present time is a serious crime due to detrimental effects on the economy of the country and hazardous health effects [22]. Forensic testing of adulteration in saffron is also a challenging task to the forensic scientist with regard to ascertain the extent of adulteration.

Adulteration of saffron with substances like beet, pomegranate fibers, and red-dyed silk fibers are sometimes observed for lowering the price and the stamens of saffron are combined with its stigma or powder to enhance the product heap. Occasionally, the flowers of different plants, especially safflower (*Carthamus tinctorious*), *Calendula officinalis*, marigold, turmeric, paprika *etc.* are deceptively added in the real stigmas. It is mentioned that adulterants (turmeric, paprika) are mixed with calcium sulfate and subsequently attached to glucose which yield 40% ash on ignition [20, 23]. The mislabeling of turmeric as "Indian saffron," "American saffron," or "Mexican saffron," also promotes fraud. So, the consumers should be careful of packets labeling above because neither of them belong to *C. sativus*. Artificial discoloration in dried stigmas to confuse the customer or adding extraneous material in the aqueous extract are the common practices and all such types of adulteration render Saffron ineffective or even unsafe medicinally for therapeutic use [24].

ANALYSIS OF ADULTERATION IN SAFFRON

First of all, a test sample is prepared. Then, various analytical methods can be used to detect adulteration in saffron, as given below:

Preparation of Test Sample

The sample is prepared as per the method given in 4 of IS 5453 (Part 2): 1996. The minimum mass of the sample is 10 g for both the whole saffron or for saffron in powder form. The above amount is enough, even if the test is to be repeated. However, if additional tests (total nitrogen and crude fibre content) are desired, a larger sample will be required.

Preliminary Examination

The methodology adopted for Forensic analysis of the saffron are according to ISO (IEC) 3632; (2) 2010.

i. **Physical Examination:** The physical examination includes testing of color, condition and texture of the saffron thread for genuinity [25]. The parameters for physical examination of Saffron are given in Table **2**.

Table 2. Physical examination of saffron.

Saffron	Color	Condition	Texture
Genuine	Crimson Red	Dry	Smooth
Fake	Crimson Red	Moist	Smooth

ii. **Color Test:** Genuine saffron imparts pale yellow color in water, while as fake saffron gives yellow coloration quickly in water.

iii. **Cotton Color Test:** Genuine saffron imparts pale yellow-orange color, whereas fake saffron yields yellow color in cotton soaked with polar solvent like water, methanol *etc.*

iv. **Whatman Paper Test:** Rubbing the fake saffron thread between whatman paper no.1 soaked with polar solvents like water *etc.* yields yellow color quickly as compared to genuine saffron.

v. **Flotation Test:** The genuine saffron floats in water while the fake saffron sinks.

vi. **Chemical Examination:** Pigments are extracted from the saffron using water, followed by water bath in watch glass to get dried pigments [5].

  a. **Reaction with Sulphuric acid:** Genuine saffron yields Indigo-blue color immediately on application of sulphuric acid to the extracted dried pigments.

  b. **Reaction with Nitric acid:** The pigment of genuine saffron imparts light blue color on reaction with nitric acid.

  c. **Reaction with Ammonia:** Pigments of genuine saffron give yellow orange color with ammonia while the fake one forms light brown color.

Molecular Methods

Molecular methods are suitable systems for tracing based on impurity in products, identified by DNA testing. DNA remains unchanged and visible in any cell, defiant to high temperature and permits species identification [26]. Application of molecular markers is being considered as an alternative for detection of safflower petals and other contaminants in commercial Saffron samples.

i. **Random Amplified Polymorphic DNA (RAPD)/Sequence Characterized Amplified Regions (SCAR):** The RAPD/SCAR method enables the obvious detection of small quantity (~1%) of adulterant so that the dubious specimens could be rejected. The SCAR markers act as quick, sensitive, dependable and cheap screening technique for the authentication of dried marketable *C. sativus* sample [27] It involves the following steps-

   1. Isolate DNA from commercial samples of saffron and safflower employing Cetyl Trimethyl
   2. Ammonium Bromide (CTAB) method [28].
   3. Perform PCR using suitable random primers in a reaction mixture of 25 µl volume.
   4. Identify amplicons which are mono-morphic for all the safflower varieties but absent in saffron samples.
   5. The putative markers amplified by the random primer, are excised from agarose gel with sterile gel slicer and purified using gel extraction kit.
   6. The A-tailed DNA is ligated into a TA-vector using rapid DNA ligation kit.
   7. The ligated vector is introduced into competent Escherichia coli strain DH5α according to the protocol of transformation by calcium chloride.
   8. The transformed colonies are picked up from the LB medium with ampicillin as selective agent.
   9. Recombinant plasmids are isolated from each overnight grown colony with high pure plasmid isolation kit.
   10. Confirmation of the clones is done by digesting the recombinant plasmid using sacI enzyme. Recombinant plasmids are sequenced by automated sequencer.
   11. Based on the sequencing, design the SCAR primers using primer 3 or other software which could amplify only safflower DNA and allows no non-specific amplification in the saffron samples in its presence.

ii. **Bar Coding Melting Curve Analysis (Bar-MCA):** It is a non-sequencing molecular method. Bar-MCA can prove to be a rapid and cheaper process for confirmatory identification of Saffron and discriminating its adulterants [16]. It utilizes chloroplast DNA bar-coding region trnH-psbA to recognize adulterants and show considerably dissimilar locations or shapes of peaks in the melting curve.

iii. **ITS Markers:** Recently, ITS markers have also been used successfully for the detection of safflower frauds in saffron [29].

Instrumentation Analysis

Extraction Techniques

Following extraction technique are used for isolation of bioactive compounds:

Solvent Based Extraction Technique

It is used for the extraction and purification of active principles. After removing fat by treating with diethyl ether, stigmas are usually extracted twice or thrice employing 70 – 90% methyl alcohol or ethyl alcohol (@10 ml. solvent /g saffron). Solvents including extract are collected, evaporated to dryness followed by purification with silica gel column chromatography employing ethanol–ethyl acetate–water (~6:3:1) as the moving phase. Crocin and crocetin can be eluted separately [30].

Steam Distillation

It is an important technique, enabling separation of volatile constituents from plants, *e.g.*, essential oils, amines, organic acids *etc.* [6]. Steam distillation can be used for detection of volatile oil constituents contributing to aroma of saffron after γ-irradiation [31].

Ultrasound-assisted Extraction (UAE)

This method uses very intense, high-frequency sound waves and their interaction with compounds. UAE is used for the recovery of biologically active glycosides (*i.e.* crocins and picrocrocin) from Saffron dry stigmas employing aq. Methanol [19, 32] and to find out the quantity of safranal by non-polar solvent extraction coupled with UV–Vis analysis for quality assessment of saffron [33].

Membrane Processes

Membrane techniques are becoming increasingly popular for the selective extraction and the fingerprinting of Saffron's constituents [34]. In this process, an emulsion liquid membrane with span 80 is used as surfactant and n-decane as membrane phase. The macroporous resin adsorption methods are employed for the isolation of crocin and picrocrocin from cell culture broth of Saffron [35]

Supercritical Fluid Extraction (SFE)

SFE is useful to recover certain principles at ambient temperature which may be subjected to thermal denaturation of the substance. Hydrolysis can be performed in a dynamic supercritical-CO2 medium for determining total safranal, *viz.* free safranal + safranal produced by picrocrocin hydrolysis [36]. The values are compared with 'safranal value' total index.

Solid Phase Extraction (SPE)

This technique is a speedy and economic one for quantitative analysis of picrocrocin by UV–Vis spectrophotometry which can be applied in the regular quality control of saffron at an industrial level [37] Glass /polypropylene columns / extraction disks are used for performing SPE [6]. To isolate the compounds of interest and improve the enrichment aspect, new sorbents have been developed. Molecularly imprinted polymer is used to determine crocin [38].

Micro-extraction

Attention is now-a-days focused on analytical procedures for extraction of constituents from complex samples and on minimization of organic solvent consumption. That is why conventional extraction techniques, *viz.* system miniaturization and/or automation *etc.* have been modified The micro extraction techniques, *e.g.* sorbent-based micro extraction technique and solvent-based micro extraction technique along with suitable sampling methods for collecting analytes enable us to decrease the number of errors during the sample pretreatment prior to chromatography as well as to minimize the adverse effects on the surroundings and the health of laboratory personnel [39], *e.g.,* Stir-bar Sorptive Extraction (SBSE) provides quantitative information regarding polycyclic aromatic hydrocarbons (PAHs) contamination in saffron [40] and ultrasound-assisted extraction (UAE) along with dispersive liquid–liquid microextraction (DLLME) method for separation and enhancement of saffron volatiles [41]. Gas chromatography–mass spectrometry (GC–MS) technique is used for separation and quantitative determination of extracted saffron constituents.

Chromatographic Techniques

Thin Layer Chromatography

A glass plate pre-coated with silica gel G is developed in chamber containing mobile phase (n-butanol, acetic acid and water in ratio of 4:1:5) [42]. For analysis, each lane receives 1µl spot of methanolic extract of the saffron. After 2 hour development, spots are visualized in daylight and UV chamber. The genuine saffron produces continuous series of yellow spots [5], while as the fake saffron gives rise to colored spots.

Gas Chromatography

A quantitative determination of Safranal can be achieved using gas chromatography. In this technique, an isothermal is run at 150°C on an SE-30 column using nitrogen as a carrier gas with a flow-rate of 30 ml/min. resulting into resolution of Safranal into sharp single peak at a retention time of 3.6 minutes [43].

Gas Chromatography Mass Spectroscopy (GC–MS)

GC–MS can be used to identify the volatile components in saffron. The quantity of volatile constituents in saffron is determined employing Safranal as a standard. After detection of odour, the lever is slided over a potentiometer to detect the deflection which in turn reflects the intensity of odour. The peaks thus obtained are transferred to the software, from the flame ionization detector (FID) to find out the peaks indicating aroma descriptors [44].

High Performance Liquid Chromatography (HPLC)

HPLC can be used to separate bioactive compounds (Crocin, crocetins and picrocrocin). A standard graph of area *vs.* concentration can be prepared comparing the results obtained with standards [43]. The compounds are separated on a gradient run from 20 to 80% (v/v) acetonitrile in water at a flow rate of 0.5 ml/ min. in 20 minutes at 308 nm.

High Performance Thin Layer Chromatography (HPTLC)

Silica gel plates (20 x 20 cm.) are used for HPTLC using solvent system- ethyl acetate: isopropanol: water (65:25:10). Yellowish spots are observed for pure/ fake saffron at different Rf values. Generally, pure saffron has lower Rf value than fake saffron. Eluted spots at above Rf values are used for U.V. spectro-photometry screening [45].

Spectroscopic Techniques

UV- Visible Spectroscopy

The maximum wavelength and the absorbance of saffron samples are determined by UV-Visible spectrophotometer. Saffron contains two glycoside components *viz.*, crocin and crocetin. Crocin and crocetin show maximum absorption at different wavelengths in different solvents, *viz.* methanol, pyridine *etc.* Crocin shows maximum absorption around 464 and 433 nm. in methanol while crocetin around 464, 436 and 411 nm. in pyridine medium.

Infrared Spectroscopy

This technique is used for detection for pure plant constituents contaminated with undesired plant materials. They can be detected by a simple, fast, sensitive, and green screening approach through IR two-dimensional correlation spectroscopy. A liquid-nitrogen cooled linear array detector with 1 x 16 elements of narrow band MCT is used to obtain visible images in a microscopic spectra using a CCD camera in the beam path. Software Spectrum IMAGE v1.7 [46] is used for the principal component analysis (PCA) of the imaging data in the complete wave number range.

Nuclear Magnetic Resonance (NMR)

Recently, the scientists have used NMR technique to produce a 'fingerprint' of the principles present in Saffron that can be compared with an unadulterated sample of Saffron. Also, this approach is sensitive, rapid, convenient and reproducible and thus can be used for screening Saffron at pilot scale. The NMR technique can be successfully used to detect adulterants such as safflower, gardenia, *C sativus* stamens and turmeric in the saffron sample when present up to a minimum level of 20 per cent by weight [47]. The 1H spectrum reveals typical constituents of saffron, *viz.*, crocin and picrocrocin along with other compounds like lipids and sugars. This forms the basis for identification of adulterants in saffron.

Microscopy

The stigmas appear either free or united in a group of three at the top of yellowish styles under microscopic examination of soaked drug. The length of each stigma is approx. 25-mm and appears as a slender, funnel shaped with dentate or fimbricate rim [48].

Electronic Nose System (ENS)

An electronic nose system can be used to generate aroma fingerprints of genuine saffron and impure samples. The different parts of ENS include MOS sensors, sample container, three electronic valves, pumps, and oxygen container. An electronic nose can detect a variety of adulterants in Saffron, *viz.*, safflower, dyed corn stigma *etc.* using six metal oxide semiconductor sensors [49].

CONCLUSION

Saffron is used in food industry as flavouring agent and as medicinal agent in various preparations for treating a variety of ailments. Adulteration of saffron with other low quality materials (Safflower, Turmeric, Paprika, Marigold, Beet, Pomegranate and red dyed silk fibres *etc.*) is now-a-days a common practice owing to its low production, expensiveness and ever increasing need in the market. Therefore, this is essential to detect adulteration in saffron for consumer's safety and healthcare. The saffron samples are analysed using physical, chemical and instrumentation methods in order to find the rapid authenticity testing of the genuine saffron.

CONSENT FOR PUBLICATION

Not applicable.

CONFLICT OF INTEREST

The author (editor) declares no conflict of interest, financial or otherwise.

ACKNOWLEDGNEMT

Declared none.

REFERENCES

[1]   Maria GC, Antonella C. Looking for Saffron (Crocus sativus L.) Parents. Funct Plant Sci Biotechnol 2010; 4(2): 1-14.

[2]   Schormuller J, Handbuch DL. Springer-Verlag, Berlin, 1970, Vol. 6.

[3]   Tarantilis PA, Polissiou M, Manfait M. Separation of picrocrocin, cis-trans-crocins and safranal of saffron using high-performance liquid chromatography with photodiode-array detection. J Chromatogr A 1994; 664(1): 55-61.
[http://dx.doi.org/10.1016/0021-9673(94)80628-4] [PMID: 8012549]

[4]   Peter W, Markus S. Saffron–Renewed Interest in an Ancient Spice. Food Rev Int 2007; 16(1): 39-59.

[5]   Shukla SK, Iqbal M. Forensic Analysis of the Saffron: Rapid Authenticity Testing. IJRASET 2015; 3(4): 228-33.

[6]   Somayeh H, Gholam HH. Extraction and Microextraction Techniques for the Determination of Compounds from Saffron. Can Chem Trans 2014; 2(2): 221-47.

[7]   Schmidt M, Betti G, Hensel A. Saffron in phytotherapy: pharmacology and clinical uses. Wien Med Wochenschr 2007; 157(13-14): 315-9.
[http://dx.doi.org/10.1007/s10354-007-0428-4] [PMID: 17704979]

[8]   Salomi MJ, Nair SC, Panikkar KR. Inhibitory effects of *Nigella sativa* and saffron (*Crocus sativus*) on chemical carcinogenesis in mice. Nutr Cancer 1991; 16(1): 67-72.
[http://dx.doi.org/10.1080/01635589109514142] [PMID: 1923908]

[9]   Nair SC, Kurumboor SK, Hasegawa JH. Saffron chemoprevention in biology and medicine: a review. Cancer Biother 1995; 10(4): 257-64.
[http://dx.doi.org/10.1089/cbr.1995.10.257] [PMID: 8590890]

[10]   Marderosian DA. Review of Natural Products, Facts and Comparison, Missouri. 2001, 520-528.

[11]   Abe K, Saito H. Effects of saffron extract and its constituent crocin on learning behaviour and long-

term potentiation. Phytother Res 2000; 14(3): 149-52.
[http://dx.doi.org/10.1002/(SICI)1099-1573(200005)14:3<149::AID-PTR665>3.0.CO;2-5] [PMID: 10815004]

[12] Saeidnia S. Future position of crocus satives as a valuable medicinal herb in phytotherapy. Pharmacognosy J 2012; (4): 71.

[13] Evans WC. Trease and Evans Pharmacognosy. 14th ed. London: WB Saunders Company Ltd. 1997; p. 438.

[14] Harper D. Online Etymology Dictionary 2001.www.etymonline.com/index.php?search=saffron

[15] Rios JL, Recio MC, Giner RM, Manez SA. Update review of saffron and its active constituents. Phytotherapy Res 1996; (10): 189−93.
[http://dx.doi.org/10.1002/(SICI)1099-1573(199605)10:3<189::AID-PTR754>3.0.CO;2-C]

[16] Jiang C, Cao L, Yuan Y, Chen M, Jin Y, Huang L. Barcoding melting curve analysis for rapid, sensitive, and discriminating authentication of saffron (*Crocus sativus* L.) from its adulterants. BioMed Res Int 2014; 2014: 809037.
[http://dx.doi.org/10.1155/2014/809037] [PMID: 25548775]

[17] John PM, Sunan W, Massimo FM. Chemical and biological properties of the world's most expensive spice: Saffron. Food Res Int 2010; (43): 1981-9.

[18] Maryam R. Chemical and Medicinal Properties of Saffron. Bull Env Pharmacol Life Sci 2015; 4(3): 69-81.

[19] Hemmati KA. Optimization of effective parameters on production of food color from Saffron petals. Agr Sci Tech 2001; (15): 13-20.

[20] Kafi M. Saffron: Production technology and manufacture. Mashhad: Ferdowsi University Publication 2002.

[21] Rastkari N, Razzaghi N, Afarin L, Alemi R, Ahmadkhaniha R. Effect of gamma irradiation on chemical properties of Saffron pigments. Acta Hortic 2007; (739): 451-3. [ISHS].
[http://dx.doi.org/10.17660/ActaHortic.2007.739.59]

[22] Iqbal M, Shukla SK, Wani S. Rapid Detection of Adulteration in Indigenous Saffron of Kashmir Valley, India. Res J Forensic Sci 2015; 3(3): 7-11.

[23] Hagh NS, Keifi N. Saffron and various fraud mannersin its production and trades. Acta Hortic 2007; (739): 411-6. [ISHS].
[http://dx.doi.org/10.17660/ActaHortic.2007.739.54]

[24] Gohari AR, Saeidnia S, Mahmoodabadi MK. An overview on saffron, phytochemicals and medicinal properties. Phcog Rev 2013; (7): 61-66.
[http://dx.doi.org/10.4103/0973-7847.112850]

[25] Srivastava TN, Rajasekharan S, Badola DP, Shah DC. Important medicinal plants of jammu and kashmir I. Kesar (saffron). Anc Sci Life 1985; 5(1): 68-73.
[PMID: 22557503]

[26] Perez J, Garcia-Vazquez E. Genetic identification of nine hake species for detection of commercial fraud. J Food Prot 2004; 67(12): 2792-6.
[http://dx.doi.org/10.4315/0362-028X-67.12.2792] [PMID: 15633688]

[27] Javanmardi N, Abdolreza B, Nasrin M, Ahmad S, Abbas HK. Identification of Safflower as a fraud in commercial Saffron using RAPD/SCAR marker. Journal of Cell and Molecular Res 2011; 3(1): 31-7.

[28] Saghai MMA, Soliman KM, Jorgensen RA, Allard RW. Ribosomal DNA sepacer length polymorphism in barley: Mendelian inheritance, chromosomal location, and population dynamics. Proceedings of the National Academy of Sciences of the United States of America. 8014-19.

[29] Babaei S, Majid T, Masoud B. Developing an SCAR and ITS reliable multiplex PCR-based assay for

safflower adulterant detection in saffron samples. Food Control 2014; (35): 323-8.
[http://dx.doi.org/10.1016/j.foodcont.2013.07.019]

[30]　Sugiura M, Shoyama Y, Saito H, Abe K. Crocin (crocetin di-gentiobiose ester) prevents the inhibitory effect of ethanol on long-term potentiation in the dentate gyrus *in vivo*. J Pharm Exp Ther 1994; (271): 703-07.

[31]　Zareena AV, Variyar PS, Gholap AS, Bongirwar DR. Chemical investigation of gamma-irradiated saffron (Crocus sativus L.). J Agric Food Chem 2001; 49(2): 687-91.
[http://dx.doi.org/10.1021/jf000922l] [PMID: 11262013]

[32]　Kyriakoudi A, Chrysanthou A, Mantzouridou F, Tsimidou MZ. Revisiting extraction of bioactive apocarotenoids from Crocus sativus L. dry stigmas (saffron). Anal Chim Acta 2012; 755(75): 77-85.
[http://dx.doi.org/10.1016/j.aca.2012.10.016] [PMID: 23146397]

[33]　Maggi L, Sanchez AM, Carmona M, *et al.* Rapid determination of safranal in the quality control of saffron spice (Crocus sativus L.). Food Chem 2011; (127): 369-73.
[http://dx.doi.org/10.1016/j.foodchem.2011.01.028]

[34]　Mokhtari B, Pourabdollah K. Extraction of saffron ingredients and its fingerprinting by nano emulsion membranes. Indian J Chem Technol 2013; (20): 222-8.

[35]　Yuan L, Huang J, Sun Z. Separation of Crocin and Picrocrocin from Saffron Cell Culture Broth by Macroporous Resin Adsorption. Food Sci 2011; •••: 24.

[36]　Zougagh M, Ríos A, Valcárcel M. Determination of total safranal by *in situ* acid hydrolysis in supercritical fluid media: Application to the quality control of commercial saffron. Anal Chim Acta 2006; 578(2): 117-21.
[http://dx.doi.org/10.1016/j.aca.2006.06.064] [PMID: 17723702]

[37]　Sanchez AM, Carmona M, Campo P, Alonso GL. Solid-phase extraction for picrocrocin determination in the quality control of saffron spice (*Crocus sativus L.*). Food Chem 2009; (116): 792-8.
[http://dx.doi.org/10.1016/j.foodchem.2009.03.039]

[38]　Mohajeri SA, Hosseinzadeh H, Keyhanfar F, Aghamohammadian J. Extraction of crocin from saffron (Crocus sativus) using molecularly imprinted polymer solid-phase extraction. Clin Biochem 2011; (44): S136.
[http://dx.doi.org/10.1016/j.clinbiochem.2011.08.326]

[39]　Jacek N, Agata S, Lukasz M. Green Sample Preparation Techniques for Chromatographic Determination of Small Organic Compounds. Int J Chem Eng Appl 2015; 6(3): 215-9.
[http://dx.doi.org/10.7763/IJCEA.2015.V6.484]

[40]　Maggi L, Carmona M, Zalacain A, *et al.* Saffron as environmental biomarker of diffuse contamination ishs. III International Symposium on Saffron: Forthcoming Challenges in Cultivation, Research and Economics.

[41]　Priscila DCC, Garde CT, Sanchez AM, Maggi L, Carmona M, Alonso GL. Determination of free amino acids and ammonium ion in saffron (*Crocus sativus L.*) from different geographical origins. Food Chem 2009; (114): 1542-8.
[http://dx.doi.org/10.1016/j.foodchem.2008.11.034]

[42]　Arya V, Thakur R. Organoleptic and microscopic analysis of Gentiana regeliana. Journal of Pharmocognosy and photochem 2012; 1(2): 31-35.

[43]　Sujata V, Ravishankar GA, Venkataraman LV. Methods for the analysis of the saffron metabolites, crocin, crocetins, picrocrocin and Safranal for the determination of the quality of the spice using thin-layer chromatography, high-performance liquid chromatography and gas chromatography. J Chromatogr A 1992; (624): 497-502.
[http://dx.doi.org/10.1016/0021-9673(92)85699-T]

[44]　Maggi L, Manuel C, Amaya Z, *et al.* Changes in saffron volatile profile according to its storage time. Food Res Int 2010; (43): 1329-34.

[http://dx.doi.org/10.1016/j.foodres.2010.03.025]

[45]   Bakre SM, Krishnamurthu R, Shinde BM. Study of Genuinity of Saffron Samples- A Case Study. J Sci Ind Res (India) 2000; (59): 596-8.

[46]   Chen J, Qun Z, Su-qin S. Adulteration screening of botanical materials by a sensitive and model-free approach using infrared spectroscopic imaging and two-dimensional correlation infrared spectroscopy. Journal of Mol. Str 2015: 1-6.

[47]   Taylor P. NMR analysis helps identify saffron adulteration. Securing Pharma - Food & Beverage 2015; pp. 1-2.

[48]   Evans WC. Trease and Evans Pharmacognosy. China: Saunders© Elsevier Limited 1996; pp. 438.

[49]   Heidarbeigi K, Seyed SM, Amin F, Mahdi GV, Shahin R, Karamatollah R. Detection of Adulteration in Saffron Samples Using Electronic Nose. Int J Food Prop 2015; (18): 1391-401. [http://dx.doi.org/10.1080/10942912.2014.915850]

SUBJECT INDEX

A

Acetildenafil 102, 104, 105, 125
Acetonitrile 39, 41, 49, 54, 59, 70, 71, 131, 169
Acid 7, 8, 41, 43, 54, 57, 70, 93, 94, 95, 96, 105, 144, 165
  chlorogenic 69
  D-Isocitric 93, 94
  ellagic 69
  formic 41, 54, 70
  linolenic 142
  malic 93, 95
  nitric 165
  salicylic 7, 8, 43
  sulphuric 8, 105, 165
  tartaric 95, 96
Acid orange 44, 53
Adulterants, common 111, 112
Adulterants in herbal medicine 130, 131
Adulteration 1, 5, 6, 7, 24. 67, 75. 83, 98
  deliberate 24, 123
  detecting 75, 98
  economic 1, 5, 67
  hazardous 1
  investigated 83
  prohibiting 6
  stop 7
Adulteration analysis methods 9
Adulteration determination 69, 70, 132
Adulteration determination of herbal food supplements 128
Adulteration identification 72
Adulteration indices 96
Adulteration level 74
Adulteration methods 91
Adulteration prohibition 65
Aesthetic adulteration 1, 6
Alkaloids 126
All ion fragmentation (AIF) 38
Aloin content 84
Amino acids 68, 75, 91, 95, 96, 97, 99

Aminocarminic acid 50, 51, 52
Amlodipine 124, 125
Amount of trans fat 145, 146, 151
Anabolic steroids 101, 111
Analysis of adulteration in saffron 164
Analytical properties 93, 94
Analytical techniques 75, 104, 140, 143, 144, 148, 161
Anthocyanins 37, 91, 92, 94, 96, 97
Antibacterial action 67
Antidepressants 109, 111
Aphrodisiac 80, 85
Argentinean cereal foods 152
Ash, soluble 163, 164
Atmospheric pressure chemical ionization (APCI) 132
Attenuated total reflection (ATR) 91
Auramine 57
Austrian pastries 145, 146
Authentication 97, 98, 128, 129, 166
Authenticity 14, 15, 73, 91, 93, 162
Azorubine 44, 57, 60, 61

B

Biloba 87, 130
Binary milk mixtures 19, 21
Binary mixes of goat and sheep milks 18
Blue and white capsule 108
Borax 7, 8
Bright blue capsules 108
Brilliant Blue FCF 39, 42, 44, 53, 58, 60
Buffalo milk 13, 21, 22, 26, 28

C

C-4 sugar content 72
Canola oil 150, 151
Carbon SIRA 95
Carmine, Indigo 39, 40, 58
Carminic acid 37, 50, 51, 52
  structures of 50

High resolution mass spectrometry (HRMS) 37, 38, 40, 54, 55, 56, 60, 61, 132

High resolution melting (HRM) 21, 31, 32

HMF content 68

Honey 65, 67, 68, 69, 70, 72, 73, 74, 75
  acacia 69, 70
  adulterated 69, 72, 74
  commercial 72
  composition of 65, 68, 75
  high cost 69
  low cost 69
  monofloral 67
  pure 67, 73, 74
  rape 69, 70

Honey adulterants 72

Honey adulteration 65, 69, 72, 73

Honeydew honey 68

Honey products 72

Hoodia gordonii 124

Hydrogen atoms 141

Hydrophilic interaction chromatography (HILIC) 130

I

Illegal adulteration 112

Illegal dye adulteration 37

Insecticides 111, 127, 128, 134

International diabetes federation (IDF) 86

International multidimensional authenticity specifications (IMAS) 91, 95

Ion mobility spectrometry (IMS) 106

Ischemic heart disease (IHD) 143

Isotope mass 46, 47

Isotope ratio mass spectrometer (IRMS) 71

Isotopes 46, 47, 73, 98
  stable 46, 47

J

Jaggery syrup adulterations 74

Joint FAO/WHO expert committee 52, 53

Juice adulteration 96, 97, 99
  monitoring pomegranate 99

L

Lambert-beer law 80, 82

Least squares, partial 74, 75, 145

Lidocaine 87

Light green SF 42, 45

Limit of detection 13, 18, 19, 20, 21

Limit of quantitation 13, 20, 27, 39

Linear dynamic range (LDR) 13, 20, 21, 27, 28, 31

Liquid chromatography 38, 65, 69, 72, 75, 91, 96, 97, 129
  high-performance 38, 91

Liquid chromatography (LC) 38, 65, 69, 70, 71, 72, 75, 91, 96, 97, 105, 129, 130, 132, 147, 169

Liquid chromatography methods for adulteration determination in honey 70

L-Malic acid 93, 94

Long chain fatty acids (LCFAs) 141

Loopholes in food adulteration act 66

Low-density lipoprotein (LDL) 92, 141

Low quality products 122

Low quality spices 3

M

Margarines 145, 146, 147

Marketable food industries 142

Mass error 49

Mass spectra of fragments 43, 58

Mass spectrometer (MS) 39, 41, 45, 46, 54, 55, 58, 59, 71, 129, 130, 131, 132, 145, 148

Mass spectrometry 15, 37, 38, 97, 106, 129

Mass spectrum 48, 49, 60, 145

Maximum absorption 169, 170

Medicinal properties 84, 161

Medium chain triglyceride (MCT) 141

Mid Infrared (MIR) 75

Milk composition 22, 29

Milks 15, 17, 21, 22, 23, 28, 29, 31
  bovine 21, 22, 28, 29, 31
  caprine 23, 31
  skimmed 15, 17

www.ingramcontent.com/pod-product-compliance
Lightning Source LLC
Chambersburg PA
CBHW041702210326
41598CB00007B/497